HADRONIC PRESS

The Early Universe

V. B. Johri

Editor

Monteroduni Castle

Campobasso Province

Molise Region

Isernia Province

ISTITUTO PER LA RICERCA DI BASE, ITALY

Series on New Frontiers in Advanced Physics
ISTITUTO PER LA RICERCA DI BASE
Castello Principe Pignatelli,
86075 Monteroduni (IS), Molise, Italy

The Early Universe

Edited by

V. B. Johri
Department of Mathematics
Indian Institute of Technology
Madras 600 036, India

HADRONIC PRESS

35246 US 19 North # 115, Palm Harbor, FL 34684 USA

U. S. Library of Congress Cataloging–in–Publication data:

Johri, V. B.

The Early Universe

Bibliography

Index

Series of New Frontiers in Advanced Physics
of the Istituto per la Ricerca di Base
Monteroduni (IS), Italy

Additional data supplied on request

ISBN 1–57485–011–3

CONTENTS

Cosmic Microwave Background Anisotropies

Ravi Subrahmanyan
Raman Research Institute,
C.V. Raman Avenue, Bangalore - 560 080, India.

Abstract

Anisotropies in the Cosmic Microwave Background are a probe of the spectrum of density perturbations in the early universe and of the astrophysical and dynamical evolution that caused the formation of galaxies and the present day observed structure in the universe. Theory of the coupling of the perturbations to the radiation predicts characteristic features in the spectrum of the anisotropies. Experiments measure the anisotropy power as viewed through spectral windows. We review the current status of the predictions of the anisotropy power spectrum and the constraints on the spectrum from observations.

Cosmology and structure formation

Observations of our universe suggest that the distribution of matter on large scales is homogeneous and isotropic. The best descriptor we have for gravity is the general theory of relativity. These together appear to suggest that the evolution of the universe follows a Friedmann model. The observation that distant galaxies show redshifts indicates that the cosmological scale factor $a(t)$ increases with time, and this cosmological expansion implies that the temperature and density of the universe were higher in the past. These arguments led to the hypothesis of the Hot Big Bang cosmology.

The subsequent realization that the light elements –^{4}He, D, ^{3}He, ^{7}Li – were possibly not formed in stellar nucleosynthesis led to the postulate that they were formed in the early universe in the hot, dense phase. Computations of the element production rates in the early universe suggested that densities of order 10^{18} cm^{-3} were necessary when the temperatures were of order 10^9 K in order to result in the observed abundances of light elements with respect to Hydrogen. Cosmological expansion to the present epoch and the present-day observed mean matter density in the universe imply a present-day temperature of about 3 K, and a confirmation of this value by the discovery of the Cosmic Microwave Background (CMB) was a triumph and validation of the Hot Big Bang Model. This is today the best theory we have for the structure and thermal evolution of the universe.

Structure formation is believed to have been seeded by density perturbations in the early universe with small fractional amplitudes. Regions of the universe with higher density expand less and regions with relatively lower density expand relatively faster, and this differential expansion results in a growth of the density perturbations. While the fractional amplitudes are small compared to unity, the perturbations grow as approximately the cosmological scale factor. When the fractional amplitudes of the perturbations exceed roughly unity, they turn-over and collapse behaving like closed universe models. The overdense regions are expected to have virialized and relaxed to bound structures that have separated from the cosmological expansion. The dynamical evolution may also be influenced by dissipative processes: astrophysical evolution. To summarize, a primordial perturbation spectrum evolves through dynamical and astrophysical processes to form the present-day observed galaxies and large scale structures: the galaxy distribution.

The coupling of density perturbations to the radiation at the epoch of decoupling, at redshift $z \approx 1100$, results in primary perturbations in the CMB. Secondary perturbations are impressed upon the CMB during the propagation of the radiation from the last scattering surface to the present epoch. Anisotropies are inevitable in a causal universe where structure forms by the gravitational growth of seed perturbations. Observations of the CMB anisotropies are a probe of the structure formation process.

Motion of observer with respect to CMB frame

The largest amplitude CMB anisotropy arises because of the motion of the observer–for example, the local standard of rest or the rest frame of the local group of galaxies–with respect to the CMB rest frame. A peculiar velocity v results in a dipole anisotropy of amplitude $\Delta T/T \approx v/c$, a quadrupole anisotropy of amplitude $Q = v^2/2c^2$ and smaller higher order terms. This effect is often referred to as the Compton-Getting effect.

Primary CMB anisotropies

The density perturbations $\Delta\rho/\rho$ at the last scattering surface result in a fluctuation gravitational potential at this surface: $\Delta\phi$. CMB photons reaching us from different points on this surface would be gravitationally redshifted by different amounts and this effect will appear as CMB temperature anisotropy on the sky.

$$\frac{\Delta T}{T} = \frac{\Delta\phi}{c^2} = (\Delta\rho/\rho)\,(r/ct_s)^2. \tag{1}$$

r is the comoving scale of the perturbation and ct_S is the comoving Hubble scale at the last scattering surface, this quantity has a value about $60h^{-1}$ Mpc. This source of anisotropy is referred to as the Sachs-Wolfe effect.

Because the fluctuations are expected to be isentropic, the matter fluctuations may be accompanied by CMB temperature inhomogenieties, and these result in a temperature anisotropy:

$$\frac{\Delta T}{T} = \frac{1}{3}\frac{\Delta\rho}{\rho}. \tag{2}$$

These temperature fluctuations partially cancel the anisotropy that arises from the Sachs-Wolfe effect.

The third important source of primary CMB anisotropies are the peculiar velocities induced in baryonic matter due to the density perturbations at the last scattering surface:

$$\frac{\Delta T}{T} = \frac{v}{c} = (\Delta\rho/\rho)\,(r/ct_s). \tag{3}$$

The total contribution due to peculiar velocities is obtained by an integral along the line of sight (through the recombination epoch) of the peculiar velocity times the probability of scattering at the epoch.

The above three primary couplings between the matter perturbations and temperature anisotropies allow CMB anisotropy measurements to be a probe of the seed matter fluctuations at the last scattering surface.

Secondary CMB anisotropies

Secondary CMB anisotropies are imprinted during the propagation of the CMB photons from the last scattering surface. Growing density perturbations during the post–recombination era–during the late non-linear evolution in a universe with critical density–will result in time-varying gravitational potentials in the medium through which the CMB is propagating. This results in CMB anisotropies because the CMB photons experience a net gravitational redshift while propagating through these potentials.

If the universe were reionized, the CMB photons are again scattered by the electrons of the medium. Sunyaev-Zeldovich scattering of the CMB by hot gas in deep potentials of clusters of galaxies will cause negative fluctuations in the sky CMB temperature (as measured in the Rayleigh-Jeans part of the CMB spectrum) along lines of sight passing through clusters. Scattering off bulk flows at the new last scattering surface (Vishniac effect) will also induce CMB anisotropies. Although the first order effect, the linear Doppler effect of order v/c may cancel, the second order effect of fractional magnitude $v\delta/c$ will survive and can contribute a significant anisotropy, as compared to the primary anisotropies, on small angular scales.

Gravitational lensing by compact objects in the universe will cause a phase mixing of CMB photon trajectories and can, in principle, generate CMB anisotropy on small angular scales.

Erasure of CMB anisotropies

During the propagation of the CMB through the recombination epoch and post recombination, anisotropies are not only generated but can also be erased by interaction with matter. The fact that recombination is not instantaneous, but takes place in quasi-equilibrium conditions over a finite time results in the erasure of small angle anisotropies whose linear dimensions are less than about the thickness of the last scattering surface. The redshift range of recombination is about $\Delta z \approx 80$ and this corresponds to a comoving length scale of about $6.6h^{-1}\Omega_0^{-1/2}$ Mpc. Propagation through recombination thus erases anisotropies below about 3.8 arcmin.

If the baryonic matter in the universe is re-ionized at a late epoch post-recombination, the last scattering surface is shifted to lower redshifts. Reionization at $z > 150$ will shift the last scattering surface to $z = 150$ and anisotropies on scales smaller than the horizon scale at this epoch, which corresponds to an angular scale of about 1.3 degrees, will be erased.

If the intergalactic matter at redshifts below $z \approx 150$ is ionized, the electron densities are not high enough to shift the last scattering surface to low redshifts, and so anisotropies are not erased; however, degree scale anisotropies are attenuated.

Describing the anisotropies

The CMB temperature is observed as a scalar function of sky coordinates on the celestial sphere: $T(\theta, \varphi)$. The anisotropies are described as fractional deviations from the mean: $\Delta T/T = (T(\theta, \varphi) - T_o)/T_o$, where T_0 is the mean CMB temperature.

The fractional temperature fluctuations on the celestial sphere may be expanded in spherical harmonics as a series expansion in terms of the $Y_{l,m}(\theta, \varphi)$. The coefficients in the expansion, $a_{l,m}$, or the rotationally invariant coefficients,

$$C_l = \langle |a_{l,m}| \rangle, \tag{4}$$

may be viewed as a power spectrum of the CMB anisotropy. The spectrum is expressed over l which is the order of the multipole. $l = 1$ corresponds to the dipole and $l = 2$ corresponds to the quadrupole. Corresponding to any multipole l is an angular scale $\theta = 60/l$ degrees and one may express the CMB anisotropy spectrum over angular scale θ instead of multipole order l. We may also transform to comoving wave numbers at the last scattering surface (at the recombination epoch): $k = l/(6000h^{-1}\ \text{Mpc})$.

To summarise, the CMB power spectrum corresponds to the rotationally invariant coefficients in an expansion of the fractional temperature anisotropy, and the spectrum may be written over either of l or θ or k.

The expected radiation power spectrum

The expected radiation power spectrum may be computed numerically by evolving the coupled Boltzmann transport equations for the radiation and matter perturbations through recombination. The post-recombination spectrum of CMB anisotropies will depend on the form of the assumed matter perturbations. Assuming a standard CDM model with a scale invariant matter fluctuation spectrum, the CMB anisotropy on large angular scales will also have a flat spectral form because the anisotropy on these scales (of tens of degrees) is primary due to the Sachs-Wolfe effect and scale invariance implies invariance in the gravitational potential perturbation amplitude with comoving scale.

On degree scales the radiation spectrum rises to give a series of peaks often referred to as Doppler peaks. This regime is where the physical cause of the anisotropy is primarily a combination of Doppler velocities and due to the adiabatic nature of the initial perturbations. Perturbations that enter the horizon prior to the recombination epoch oscillate as sound waves and undergo successive transformations between being pressure fluctuations and velocity fluctuations. The amplitude of the CMB anisotropy on a specific comoving scale depends on the state of the fluctuation on that scale at the recombination epoch. Since the state is an oscillatory function, the CMB anisotropy power spectrum will also be an oscillatory function of the multipole order. The relative height of these Doppler peaks with respect to the flat spectrum on larger angular scales depends on the baryon density of the universe.

On the angular scales well below a degree, the spectral power in CMB anisotropies is progressively reduced. The cut off is due to (a) the finite thickness of the last scattering surface as mentioned earlier, and (b) due to the velocity damping of perturbations that have entered the horizon in the pre-recombination era.

To summarize, standard theory of structure formation expects a CMB anisotropy spectrum of the following form: A flat spectrum on large angular scales of several degrees, a series of successive peaks on scales just short of a degree, and an exponential cutoff on smaller arcmin scales. Expected CMB power spectra in a variety of models of structure formation are reviewed in White *et al.* (1994).

Measurement of CMB anisotropy

The first definite measurement of CMB anisotropies was made on scales exceeding 7 degrees by the COBE-DMR experiment, which was a satellite based differential radiometer measurement. The COBE experiment produced all-sky images at multiple frequencies–31.5, 53 and 90 GHz–and detected a sky temperature fluctuation above the expected instrument noise. The COBE measured anisotropies on multipoles $l \leq 10$ which correspond to comoving scales exceeding $600h^{-1}$ Mpc. COBE-DMR measured the dipole anisotropy with amplitude 3.343 ± 0.016 mK corresponding to a velocity of the local group of about 627 km s^{-1}. The quadrupole amplitude was 6 mK. With the dipole subtracted, the residual image of the CMB sky had an excess rms of 30 μK at a resolution of 10 degrees.

At the other end of the CMB spectrum, on arcmin scales, the most sensitive measurement of the CMB anisotropy is the Australia Telescope Compact Array (ATCA) observation (Subrahmanyan et al. 1993). In contrast to the COBE-DMR observations, which were sky difference measurements with a differential radiometer, the ATCA observations used an array of antennas to measure the spatial coherence function of the electromagnetic field on the ground due to the CMB. The spatial coherence measurements are Fourier-inverted to obtain the fluctuations on a sky patch defined by the antenna primary beam pattern. The antenna diameter of 22 m and spacing of 30.4 m defines the experiment parameters: the ATCA observations are sensitive to CMB anisotropy on a narrow window centered at about $l = 4500$ corresponding to $\theta = 1$ arcmin or $k = 1\ h^{-1}$ Mpc. The ATCA experiment probes CMB anisotropies that are progenitors of galactic scale structure. The ATCA experiment (1993) was a negative result and did not detect anisotropy: the experiment placed an upper limit of 25 μK on CMB fluctuations on the sky when smoothed to about 1 arcmin resolution. It may be noted that the experimental sensitivity (in temperature units) on the sky image produced by ATCA is lower than the sensitivity of the COBE images.

A large number of experiments have been attempting to measure the CMB anisotropy and each experiment views the CMB spectrum through a well defined window function that is determined by the observing scheme. The experimental techniques used in several experiments are reviewed by Readhead and Lawrence (1992). The COBE window is a low pass filter in l-space; most other experiments adopt switching schemes that make the window functions band-pass filters in l-space.

On large angular scales, $l < 30$, firm and confirmed detections have been made by the COBE and MIT and TENERIFE experiments. On intermediate scales $l \approx 10^2$, several experiments like the PYTHON, ARGO, SP91, SASTAKOON, MAX GUM and μP regions and MSAM experiments all claim detections. On small angular scales, $l > 10^3$, there are no detections as yet; the VLA, OVRO and ATCA have all made sensitive measurements in this regime without positive detections. The ATCA experiment is the most sensitive on this scale and is continuing to improve as the observations continue.

Confrontation between observations and theory

The CMB anisotropy detections on large angular scales are currently useful in providing a normalization for the amplitude of the matter perturbation spectrum. The COBE experiment is also consistent with a scale invariant form for the matter perturbation spectrum on large scales.

The COBE normalization implies an amplitude of 17 μK at $l = 2$. This corresponds to a $\Delta T/T$ of approximately 6×10^{-6} on large angular scales with a flat spectrum. A model universe in which structure forms by the initial formation of supercluster condensates which then fragment to form the galaxies we see today will require a matter perturbation exceeding $\Delta\rho/\rho$ of order unity at redshift of at least 5. Since linear evolution of density perturbations follows $\Delta\rho/\rho \sim (1+z)^{-1}$, the fractional density perturbations at $z = 1100$ would be expected to be at least (5/1100). The Sachs-Wolfe effect expects the resulting fractional anisotropy to be of order 10^{-4}. The expected CMB anisotropy in this model for structure formation is about two orders of magnitude larger than that observed, implying that top-down pancake model for structure formation is definitely ruled out by the CMB observations.

Detailed comparison of the expected CMB anisotropy in specific models of structure formation with the observations constrains the parameters of the models.

It is interesting and encouraging that the matter perturbations on large scales implied by the large angular scale CMB measurements appears to be a continuation of the matter perturbations on small scales as derived from the observed distribution of galaxies today. The degree scale measurements are also showing indications of being of higher amplitude as compared to the large angle measurements, as expected due to the Doppler peaks. The small angle measurements are also consistent with theory: they are consistent with a cutoff at large l. Continued observations will help to constrain models of structure formation better.

References

1. Readhead A.C.S. and Lawrence C.R., ARAA 1992, **30**, 653.
2. Subrahmanyan R., Ekers R.D., Sinclair M. and Silk J., MNRAS 1993, **263**, 416.
3. White M., Scott D. and Silk J., ARAA 1994, **32**, 319.

Gravitational Lens and Large Scale Structures in the Universe

D. Narasimha
Tata Institute of Fundamental Research,
Homi Bhabha Road, Bombay - 400 005.

Abstract

A point lens produces two images of a background source. The position, magnification ratios and time-delay between the images can uniquely determine the mass of the lens and the *effective distance* to the lens system. It is possible to extend this principle to probe an extended lens having sufficient projected surface mass density to form multiple images. However, to obtain a reliable estimate of the mass profile, the shear due to the ellipticity of lens mass distribution as well as focusing action of the distributed mass in the lens should be suitably modelled. In this way extended structures over the scale of parsecs (globular clusters) to a few megaparsecs (superclusters) can be probed through their gravitational lensing action, by studying at good resolution either the multiple images or the morphology of the distorted extended image of background sources along their line of sight. When the images cannot be resolved, analysis of the flux variability due to the relative motion between the lens and the background sources could provide constraints on the mass of the lens. The ongoing microlens experiments *MACHO*, *OGLE* and *EROS* have given valuable information relating to mass distribution in our Galactic disk, bulge and halo after monitoring light variability in stars in LMC and the Galactic Bulge for the last four years.

Lensing of background galaxies by rich galaxy-clusters at moderate redshifts can produce extended arc-like features due to large linear magnification. If the cluster lensing is dominantly due to the smooth mass distribution and if more than two multiply-imaged lensed galaxies could be observed near the cluster-centre, the position of the *breaks* or *mergers* in the images and relative lengths of individual images can be used to infer the large scale geometry of the Universe. Presently slight distortion of distant galaxies distributed over a field of few arcminutes due to *weak lensing* by foreground cluster is being used as one of the possible inversion techniques to obtain the cluster mass distribution. With improved observations, the technique could become a powerful probe of the gravitational mass distribution at megaparsec scale.

1. Introduction

In the last decade a new observational technique has emerged to probe the mass distribution in large scale structures in the Universe in an unconventional way, through the gravitational effect on the light rays traversing through the structure. Gravitational lensing has the promise to deliver answers to some of the basic problems that have bogged the astrophysicists and cosmologists for the last few decades, like determination of the mass distribution in our Galaxy, large scale geometry of the Universe, the Hubble Constant or estimation of the mass distribution in objects at large distance at scales of less than $10^{-4}M_\odot$ to the supercluster scale. Lensing occurs when the gravitational field of an intervening object bends and distorts the light beam from a background source - thereby altering the position, size and shape of the observed image(s). Though lens effect is a consequence of General Theory of Relativity, the weak field approximation is adequate for analysis of the lens systems observed, where the bending angle is of the order of arcseconds.

Gravitational lensing became an active observational field after Walsh *et al.* (1979) discovered the twin system 0957+561. Since then more than two dozens quasar and AGN lens systems and candidates have been reported in addition to a dozen radio rings and arc-like structures of extent of subarcsecond to arcseconds. Quasars and AGNs at high redshifts had been the choice candidates for lensing during the early day searches. If lensed by an intervening foreground galaxy situated at moderate to high redshift, we expect to observe multiple images of the quasar having identical redshift and similar spectral characteristics. Time-delay in the flux variability of the lensed AGN has been observed in three systems, though the value of time-delay in any of these cases is yet to be firmly established. The main problem in deciphering the signal from the *noise* has itself been *another important probe of high redshift galaxies*, namely, *microlensing by stars in the lens galaxy*. In almost all the galaxies which produce images of arcsecond extent, the optical depth for microlensing by stars and other compact objects in the lens is of the order of unity. Consequently, most of the time one or more of the images are susceptible to random flux variability caused by microlensing due to stars (cf. Narasimha, 1994b).

At the lower end of the mass scale, *microlensing* has provided valuable probe of mass density in star-like objects in galaxies. Compact objects like stars or brown dwarfs, passing in front of a compact background source can produce achromatic variation in the radiation flux from the source, a systematic study of which could provide reliable estimate of the amount of mass in such microlenses. At present three major experiments are progressing, to monitor microlensing of stars in our Galactic bulge and in LMC by compact objects in our galaxy. Soon after MACHO (Alcock *et al*, 1993) and EROS (Aubourg *et al*, 1993) announced their first results of search for microlensed stars in LMC, OGLE (Udalski *et al*, 1993) reported microlensing of stars in our Galactic Bulge. Since then more than 40 candidates have been analysed by the MACHO and OGLE group. A somewhat similar method based on the linear magnification and distortion of jets and knots of radio sources due to lensing could provide evidence for the existence of massive black holes, or probably constrain their number density. The Jodrell Bank astronomers have been studying this problem (Henstock *et al*, 1993; Calder *et al*, 1993).

At the larger scale, compact rich clusters of galaxies situated at redshifts upwards of 0.2 are capable of forming multiple images of background sources located near their centre because the *Mass to Light Ratio* in their core is of the order of 200 to 500 $M_\odot/L_\odot$. The background galaxies, lensed by these galaxy-clusters are imaged into elongated arc-like structures of tens of arcsecond extent. Detection of linear structures (or arcs) with or without a counter image near the cluster centre provide valuable diagnostic on the mass distribution in the galaxy-clusters. Even if no evidence for strong lens events are observable, almost any moderate to rich cluster at these redshifts exhibit signatures of *weak lensing* - faint background galaxies within a few arcminutes from the cluster centre will be systematically distorted which could be statistically inferred from the orientation of the major axes of the galaxies. Cluster lenses have been topics of intense observational and theoretical study ever since Soucail *et al.* (1987) and Lynds & Petrosian (1986) reported detection of a giant arc near the centre of the galaxy-cluster A370. More than two dozen clusters at redshifts of up to approximately 0.6 have exhibited arcs or slightly distorted galaxies in their field. Inversion for the mass distribution of the lens cluster from the shear of background lensed galaxies has been a hot topic in the last few years (Kaiser & Squire, 1993). Imaging down to very faint limits and spectroscopy of the arcs have enabled to model the mass distribution in clusters of galaxies (Mellier *et al., 1993, 1994; Narasimha & Chitre, 1988, 1993a)*. The spectroscopy of the extended arc in Abell 2390 (redshift 0.231) has enabled to obtain its approximate rotation curve (Soucail & Fort, 1991; Narasimha & Chitre, 1993b) confirming that the lensed source is a spiral galaxy at redshift of 0.914. This opens up an avenue to probe faint galaxies at high redshifts which are magnified by the foreground clusters by a factor of 10 or more.

In this presentation only the following four aspects of lensing will be outlined: microlensing as a probe of mass distribution in our Galaxy, subarcsecond lenses, cluster lenses as probe of large scale geometry of the Universe and weak lensing. The choice is somewhat arbitrary, but I believe that

Microlensing offers good prospects of diagnostic of stellar and sustellar mass at the galactic scale and definitive results about mass spectrum of stellar objects our Galaxy is expected with in a couple of years.

Polarization studies of subarcsecond lens systems, which could probably be imaged by edge-on spiral galaxies at high redshifts could be valuable probes of galaxy formation and cosmic magnetic field. It also will be a good method to measure time-delay if beam depolarization is properly taken into account.

Galaxy-clusters are the key to structure formation and probably the best bet for a glimpse of the large-scale geometry of the Universe.

The diagnostic value of lens will be illustrated in the next section through the example of a point mass. The microlens experiments to probe the mass distribution in our Galaxy and some preliminary inferences drawn from their results will be discussed in section 3. Some idea of extended thin lens will be given in section 4 and a brief discussion of subarcsecond lenses and their cosmological implications will follow in section 5. A method to determine the large scale geometry of the Universe using cluster-lenses will be the topic of section 6 and some simple theoretical aspects and limitations of weak lensing will be sketched in section 7.

2. Some Basic Ideas of Gravitational Lensing

Let a source, situated at a distance D_S from the observer be at an angular separation θ_s from a foreground point lens of mass M at a distance D_L from the observer. Let the distance between the source and lens be D_{LS} and we shall denote an effective distance D_{eff} by,

$$D_{eff} = \frac{D_L D_S}{D_{LS}} \tag{1}$$

(All the distances here are angular diameter distances). The lens equation is,

$$\theta_s = \theta_I - \frac{\mu}{\theta_I^2} \theta_I, \tag{2}$$

where,

$$\mu = \frac{4GM}{D_{eff} c^2}. \tag{3}$$

Here G is the gravitational constant and c, velocity of light. Two images of the source will be formed at angular positions given by,

$$\theta_1 = \frac{\theta_s \pm \sqrt{\theta_s^2 + 4\mu}}{2} \tag{4}$$

and the magnification of an image is given by,

$$A_I = \left[1 - \frac{\mu^2}{|\theta_I|^4}\right]^{-1}. \tag{5}$$

Intrinsic variability in the source will appear in the two images at different epochs separated by an interval (called the time-delay) given by

$$\tau = \frac{D_{eff}}{c}\left[\mu \, ln \frac{\theta_2}{\theta_1} + \frac{\theta_2^2 - \theta_1^2}{2}\right], \tag{6}$$

the signal appearing first in image 2 if $\tau > 0$. Inverting the above relations we find that

$$\mu = \frac{\Delta\,\theta^2 \sqrt{A}}{(1 + \sqrt{A})^2} \tag{7}$$

and

$$D_{eff} = 2c\tau \frac{\delta\theta^2}{(1 + \sqrt{A})^2} [\sqrt{A}\, lnA + A - 1]^{-1}, \tag{8}$$

where A is the magnification ratio and $\delta\theta$ is the separation between the images.

Even if the image separation is too small to resolve and the lens is unobserved, it is possible to probe the mass of the lens if the signature of its motion in front of the source, flux variability due to *microlensing*, could be analysed. Let θ_m be the impact parameter and the motion of the lens normal to the line of sight can be expressed as

$$\theta_s = \theta_m \vec{i} + a(t - t_0)\vec{j}, \tag{9}$$

where t and t_0 denote time, a, relative angular velocity of the source along the direction $\vec{j}$. Define

$$x = \frac{\theta_m^2}{4\mu} \tag{10}$$

and

$$y = \frac{(a(t - t_0))^2}{4\mu}. \tag{11}$$

The sum of the absolute magnification of the images can be shown to be

$$M(t) = \frac{1 + 2x + 2y}{2\sqrt{x+y}\ \sqrt{[1+x+y]}} \tag{12}$$

For an isolated point lens, the flux variation is determined by the following two parameters:

1. The impact parameter, $\sqrt{x}$, expressed in units of the strength of the lens (we have used $2\sqrt{\mu}$) and
2. The relative transverse angular velocity of the lens with respect to the source, $a/\sqrt{\mu}$ (also scaled with the strength of the lens).

We have displayed, in Fig.1 the microlens light curve due to a point lens for a range of impact parameters. The figure can be scaled if we know the mass of the lens and the distance scales as well as the angular velocity of the lens. Notice that probability for having an event of specified peak magnification drops off nearly as the inverse of the magnification for microlens events of large magnification.

We can summarise the conditions for observability of a point lens as follows (Saslaw, Narasimha & Chitre, 1985):

If a source has angular size larger than $\sqrt{\mu}$, the lens can be detected through light variation due to microlensing.

If the *Radius of Gravity Ring* or *Einstein Radius*, $\sqrt{\mu}$ is larger than the angular resolution of the instrument and is larger than the angular scale of brightness variation in the image, the lens can be observed through the multiple images of the source.

3. Microlens as a Probe of Compact Objects in Galaxies

Microlensing offers a method to study the mass distribution in a galaxy in compact objects. The peak to asymptotic luminosity of a star lensed by a point object is a direct measure of the impact parameter in the units of $\sqrt{\mu}$, the strength of the lens. The duration of the event is directly proportional to $\sqrt{\mu}$ but varies inversely with the angular velocity of the lens. Consequently, microlensing of sources of angular size substantially smaller than $\sqrt{\mu}$ can be used to estimate the mass of the lensing objects if we can rule out other possible sources of variability in the source. The departure of the light curve from the profile $M(t)$ (cf. Eq. (12)) is an indication of the finite size of the lens or source, intrinsic variability or possible multiple constituents of the lens. When an object of mass M situated at distance D from the observer lenses a source at the centre of the Galaxy, the scale over which point source approximation is valid is $\sim 10^{14}\sqrt{M/M_\odot}\sqrt{D/10kpc}$ star last for a month. The fraction of variable stars which shows comparable flux variability is $\sim 10^{-3}$. Clearly, accurate photometry in multiple wavelengths for a few years will be required to rule out intrinsic variability of the source or noise in the data.

The Macho Projects:

Three projects are in progress to detect microlens events by observing the light curves of stars in LMC, SMC and M31 and the Galactic Bulge, namely, MACHO (Alcock *et al*, 1993), EROS (Aubourg *et al*, 1993) and OGLE (Udalski *et al*, 1993). MACHO, the US-Australian collaboration, reported one event after monitoring 1.8 million stars in LMC and over 45 events after following more than 10 million stars in Galactic Bulge (Alcock *et al*, 1994). Some of the Galactic events could be intrinsic variability but it is expected that more than 95% of them could be microlenses though some could be stars in the Galactic Disk. EROS, the French collaboration monitoring the LMC stars from their ESO site at La Sila, reported two good events after analysing approx. 3 million of their sample (Aubourg *et al*, 1993). OGLE, a collaboration between Poland and US, monitors stars in the Galactic Bulge from the $1m$ Swope telescope at the Las Campanas Observatory in Chile since 1992 and reported the first Galactic Bulge microlens event in 1993 (Udalski *et al*, 1993). After analysing the light curves of 1.4 million stars in 1992 and 1.1 million stars in 1993 they have reported 9 microlens events including one double-lens event and also have captured one event form real time monitoring (Udalski *et al*, 1994).

Microlens light curves for various impact parameters are displayed for an isolated point lens. The sum of absolute magnifications of the two micro-images is plotted against the x-component of the microlens position. The x-axis as well as impact parameter are in units of $\sqrt{\mu}$.

There are two completely independent methods of analysis - for the estimation of the mass of the individual lens and the total mass in all the objects which could produce microlensing with in the sample (Narasimha 1994a,b). It should be stressed that after an microlensing event is detected, constraints on the *mass of the individual microlens can be obtained from the failure to observe microlensing of neighbouring stars or, more reliably, from the time interval between variability in neighbouring stars and their peak magnifications.* Other possible methods for mass determination by studying possible microlensed standard variable stars was

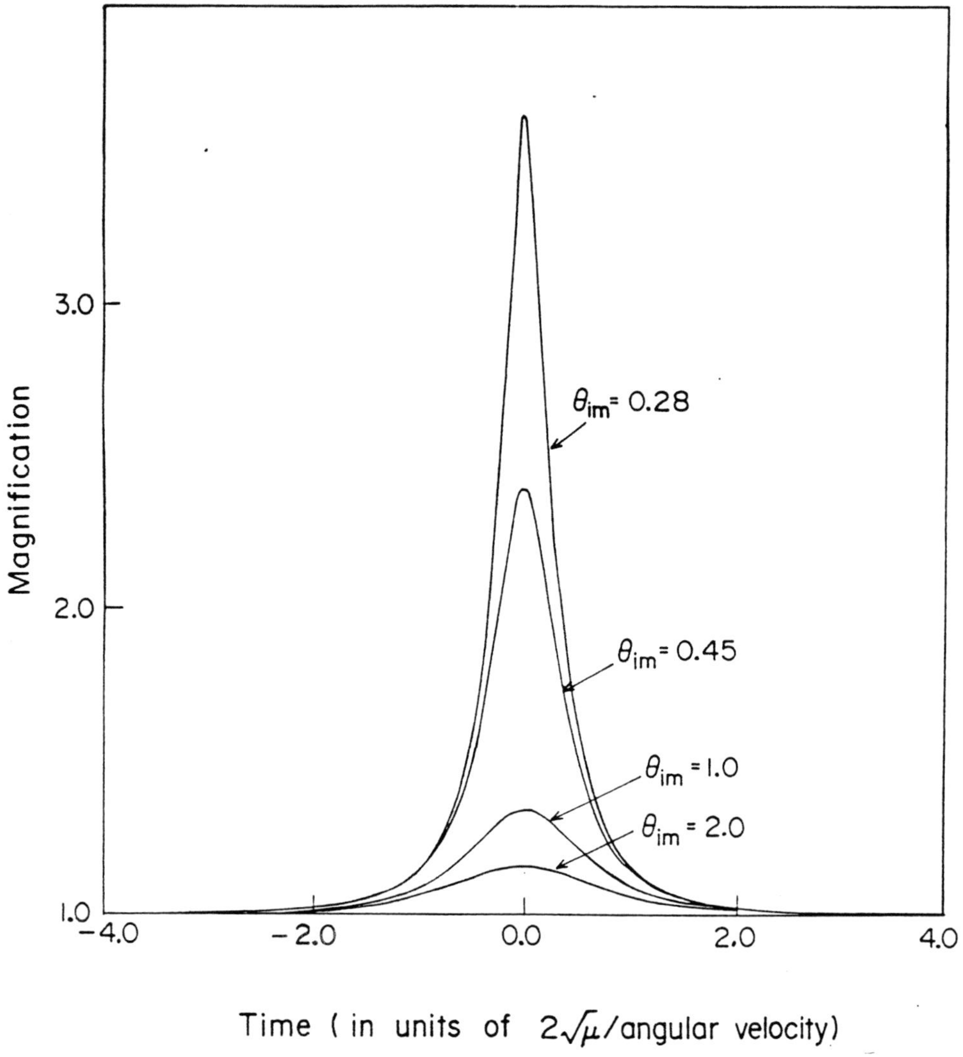

Fig. 1 Flux variability in a point source due to Microlensing

speculated by Narasimha (1994b). However, the total mass in such microlenses (optical depth for microlensing) can be estimated by studying all the microlensing events in the sample. From Eq.(12), it is apparent that there will be more numerous low magnification events corresponding to large impact parameter than the large magnification good events, which require impact parameter of the order of 0.2 in the units we have used. But the two main difficulties in the measurement of the optical depth appear to be,

1. Determination of the efficiency of a programme to detect an event of specified duration for a given type of stars.
2. Estimation of the distance to the lensed stars. Both MACHO and OGLE groups are considering the red clump stars to be more reliable compared to the standard stars. This is borne out by their results: While the optical depth estimated for standard stars (as of last month) varied by a factor of six between various analyses of the groups, for the red clumps it was with in a narrow range of $3^{+1.5}_{-0.9} \times 10^{-6}$ (Alcock *et al*, 1994) to $(3.3 \pm 1.2 \times 10^{-6})$ ((Udalski *et al*, 1994).

Inferences on the Mass Distribution in our Galaxy

It will be difficult to derive meaningful conclusions on mass distribution in the Galactic Halo based on the present observed microlenses. While the optical depth estimated is larger than the value derived from the observed luminous mass in the halo, it is smaller than the value we would expect if the maximum permitted mass of the halo is made up of compact objects like stars and brown dwarfs.

The results for the disk too are preliminary because the efficiency of detection of microlenses is still poorly determined. The OGLE and MACHO results have established that the optical depth for microlensing towards the Galactic Centre is around $(3-7) \times 10^{-6}$, which is considerably larger than the value of 1×10^{-6} we expect based on the baryonic mass distribution with in the solar neighborhood. OGLE also gives strong arguments for the existence of Galactic Bar at kiloparsec scale (Paczynski *et al* 1994). Possible existence of bar implies the following:

A massive halo is no longer necessary for the sake of stability of the Galactic Disk. However, according to the numerical simulations a disk forming bar-type of structure should also soon possess random velocity comparable to the rotational velocity. The observed random velocity in the Galactic Disk is, however, small.

The optical depth for microlensing of sources along the line of sight towards the bar will be enhanced. This possibility should be testable by monitoring various windows towards the Bulge. If indeed there be a bar at a few degrees along the line of sight with mass density of $10^5 M_\odot parsec^{-3}$, the characteristic configuration of the multiply-imaged background source will be readily discernible in, say, radio maps.

Microlensing by Compact Objects in External Galaxies:

Microlensing events have been observed in lens systems at large redshifts. In fact, very soon after its discovery, 0957+561, the first observed lens, was systematically monitored for time-delay estimation and it was found that the images show light variation independent of each other which could not be explained by any consistent time-delay. (Schild & Weekes, 1984). But the best confirmed case is probably the Huchra Lens, 2237 + 0305 (Irwin *et al*, 1989). Rauch & Blandford (1991) modelled the microlens flux variability in 2237 + 0305 to conclude that the optically thick accretion disk models of AGNs (having size of the order of 10^{16} cm) is in conflict with the lens light curve. However, Jarosynski, Wambsganss & Paczynski (1992) used α-disk model and concluded that the light curve does not contradict with the disk models.

There have also been investigations to detect possible $10^6 M\odot$ objects through milli-lensing. The principle used was one of the following:

Comparison of maps with numerical simulations: The Jodrell Bank Groups studied objects from their sample at a few to hundred milliarcsecond resolution for lens signatures (Henstock *et al*, 1993). They used the TIFR lens code to numerically simulate possible milli-lens signatures expected for their sources.

VLB transformation matrix (cf. Narasimha *et al*, 1984): Let a multiply imaged system consist of radio jet at milliarcsecond scale. If lensed by a smooth mass distribution alone, the images of the jets, though distorted differently, can be superposed one over another through a smooth transformation. However, if a blackhole or an opaque object deforms one of the images, it will introduce a knot or island type of structure and consequently, a smooth transformation of one image of the jet into the other is not possible. By monitoring a lens over a period of time, this method can be used to give limits on the number of possible massive compact objects in the lens galaxy, but in spite of prolonged monitor of 0957 + 561 the upper limit on the mass fraction of massive blackholes in the lens galaxy is 0.1 (cf. Calder *et al*, 1993).

4. Extended Lenses

A realistic lens like a galaxy cannot be approximated by a point mass. The main differences between a single point lens and an extended lens like a galaxy are two-fold:

1. For a compact opaque lens, the bending angle monotonically increases towards the centre of the lens, and photons approaching closer than the radius of the lens cannot pass through the lens. Consequently, the centre of the lens exhibits a discontinuity in the bending angle. For a smooth bounded transparent lens the bending angle resembles that of a point lens at large distances, but close to the centre the bending angle smoothly drops to zero. An additional odd image, generally highly demagnified, is formed near the lens centre where the bending angle drops off rapidly. *Detection of this odd image and determination of its flux will provide a powerful probe of the mass distribution of the nucleus of the lens galaxy (Narasimha, Subramanian & Chitre, 1986). We believe that this image has been detected in the radio loud systems like* 1830-211 *(Subrahmanyan et al, 1990),* 1422 + 231 *(Narasimha & Patnaik, 1993) and generally imply a mass of* ~ 10^9 M⊙ *within 50 to 100 parses from the lens centre.*

2. The ellipticity of the mass distribution in the lens can produce a variety of configurations due to the shear on the photon beam introduced by the mass distribution in the lens. [For a circularly symmetric projected mass density, the projected position of the source, lens and the images will be collinear. This condition restricts possible deformation of extended images]. In the simple spheroidal lens, up to seven images of a background source can form if the source is suitably placed.

For thin lenses (objects whose size is small compared to the distances between the observer, source and the lens) it is possible to obtain the solution of the equation of gravitational lens for a specified source position using the complex formalism of Bourassa and Kantowski (1975) and construct the lens models (Narasimha *et al*, 1982) or we could try to solve the two dimensional Poisson's equation for the lens potential,

$$\nabla_H^2 \Phi_L = 2\kappa. \tag{13}$$

The source term of the Poisson's equation, the *convergence*, which represents the focusing due to the matter within the beam of light is,

$$\kappa = \frac{\Sigma}{\Sigma_{cr}}, \tag{14}$$

where Σ is the surface mass density of the lens and the critical surface mass density is given by,

$$\Sigma_{cr} = \frac{c^2 D_{eff}}{4\pi G} \tag{15}$$

(We have used angular coordinates throughout our discussions) and the bending angle will be the gradient of the lens potential. The lens equation, thus, becomes

$$\theta_I - \theta_s = \nabla_H \Phi_L. \tag{16}$$

This formulation can provide some insight into the problem and possible approximations when the lens consists of components having a range of length scales like Jupiter-sized microlenses to superclusters, but we shall avoid the discussion. The lensing could be considered to be *strong*, and hence the magnification of the background source can be substantial if Σ is comparable to Σ_{cr} or greater for the specified source and lens distances. If Σ is small, there can still be some effects of lensing *weak lensing* which can only be inferred by statistical means. For a fixed angular position, κ increases with the source distance. Consequently an extended object, which acts as a weak lenses at distance close to the source, can form multiple images of a source at large enough distance.

At what scales are large scale structures observable through their lensing action? It depends on the distance scales as we noted earlier, and strong lensing is more probable at larger effective distances though observationally it is more difficult to detect sources at larger distances. Consequently, assuming that the lens and source are situated at an optimal distance for detection of the lens signatures, we can estimate the range of mass scales in the Universe which can be probed by lensing. In *Table*.1 the typical value of Σ/Σ_{cr} for various structures are given. We find that transparent lenses in the scale of globular clusters to galaxy-clusters can be probed through *strong lensing* while, it might be possible to probe superclusters through *weak lensing*.

The main diagnostic value of lens is due to the distortion of the images, which is given from the transformation matrix between the source and image planes.

$$[A] = \left[\frac{\partial\theta_s}{\partial\theta_I}\right]^{-1}. \tag{17}$$

The locus of the zeros of the determinant of this matrix $[A]^{-1}$ form the *caustics* in the image plane (where the intensity of the images become infinity) for a given mass distribution of the lens and distances to the source and lens. The corresponding points in the source plane form the *critical curves*. (The mathematicians and many astrophysicists differ from each other in terminology here). The *caustics are the singularities of the mapping from the image plane to the source plane.* By studying the critical curves, we can get an idea of the image multiplicities as well as expected distortion of various images in gravitational lens systems. The number of images changes across the critical curves and sources crossing them form extended merging arc-like features. Even a small source crossing a critical curve will form an extended magnified image. It might be useful to classify a caustic as tangential or radial depending upon whether the infinity magnification is along the caustic or normal to it. For a circular lens, (if surface mass density is above or equal to the critical mass density,) only a radial caustic is present and the tangential critical curve is a non-generic point centered at the lens and the corresponding caustic is the *Einstein Ring* or *the cone of inversion* (Saslaw, Narasimha & Chitre, 1985).

Table 1. Structures which can be probed through Gravitational Lensing
(numbers are illustrative only)

Object	Mass $M_\odot$	Scale Length kiloparsec	$\kappa = \frac{\Sigma}{\Sigma_{cr}}$	Optimum angular scale (arcsecond)
Globular Cluster* Giant Molecular Cloud*	10^6	2×10^{-3}	0.5	0.01
Galaxy	10^{10}	0.5	1	0-.5
Giant Galaxy	10^{12}	5	2	2
Dark Galactic Halo	10^{13}	20	1	5
Galaxy-cluster	10^{15}	200	0.5	10
Supercluster	10^{17}	2×10^4	0.01	10^3

* only if the host galaxy by itself is not a strong lens.

Though realistic lenses are inhomogeneous at all scales, the lens potential of the projected surface mass density is comparatively smooth at scales greater than parsecs. Consequently, the mass distribution in galaxies and galaxy-clusters can be approximated by assuming elliptical symmetry. *But the lens potential will not be elliptically symmetric.* For elliptical symmetry of mass distribution, the central value of κ, the scale-length at which it varies and its eccentricity determine the image properties. The lensing of the galaxy becomes important as the source distance increases and $\kappa(0)$ is comparable to unity. In figure 2, we have displayed the caustics and the morphology of the images for two sources at different redshifts, lensed by a galaxy-cluster which has (1) surface mass density just sufficient for multiple image formation (*marginal lensing:* Kovner, 1987; Narasimha, 1993) resulting in almost line-like lip caustic and (2) κ is higher because of the larger source distance but still only the lip caustic is formed. Though these configurations have very small cross section they have important applications for diagnostic of clusters of galaxies. The large magnification parallel to the major axis of the lens and the merging of images, which are signatures of marginal lensing, give rise to characteristic shape of the image not sensitively dependent on the shape of the source.

When $\kappa(0)$ becomes larger, through a series of deformations in the caustic the critical curve bifurcates into two lips perpendicular to each other. For higher mass densities, a diamond shaped tangential critical curve and an oval shaped radial critical curve become distinct. There are many combinations of caustics possible in a realistic lens system. Some of these curves and the corresponding signatures of lensing expected when extended sources are imaged are described by Narasimha (1994b).

The image morphology of a gravitational lens system consisting of extended sources gives an idea of the caustic structure. When any of the sources cross a critical curve, the position of the caustic becomes clear from an inspection of the image. If we are able to draw the approximate caustics and have some *a priori* idea of some of the generic properties of the source, then for a specified form of the mass density of the lens galaxy or cluster reliable model of the lens can be constructed. It should be noted that in the absence of these external *a priori* constraints, the lens models will not be unique and the inferences drawn from them will be very tentative.

5. Galaxy-Lenses

Many subarcsecond lenses have been discovered in radio surveys (cf. Patnaik, 1993). Among the dozen known radio lenses, five have separation less than or nearly one arcsecond. Some of the extensively observed among these are $1830-211, 0218+357, 1422+231$. However, $1938+666$ and $0751+272$, in spite of their intrinsic faintness, have good arc-like features and additional structures which make their models very robust. Preliminary models for most of these systems were given in Narasimha & Patnaik (1993). Detailed models incorporating the three decades of multifrequency observations of $1830-211$ are discussed in Nair (1993) and Nair, Narasimha & Ras (1993). High resolution VLBA observations showing sub-milliarcsecond structures of $0218+357$ have been obtained by Patnaik *et al* (1994) which complement the earlier MERLIN and other low resolution observations of the ring like features.

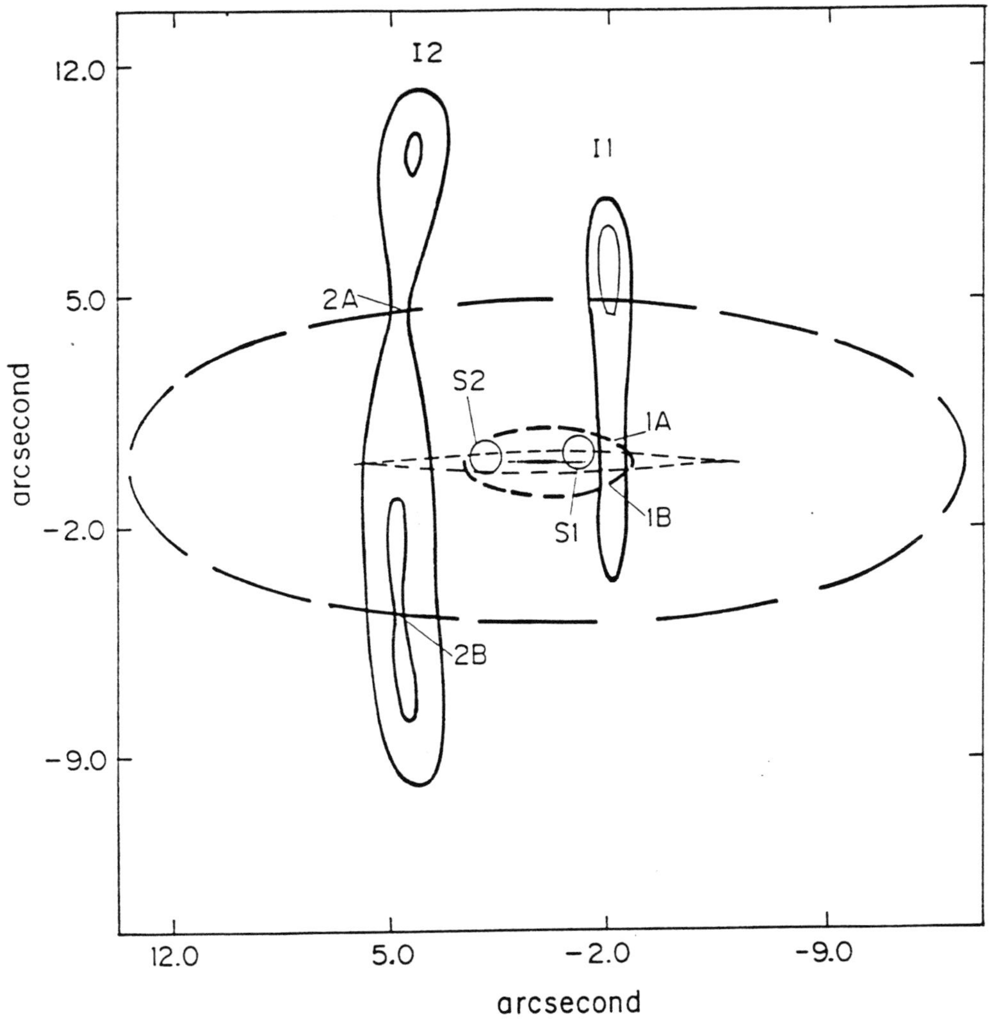

Fig. 2. Caustics and image morphology due to a smooth cluster-lens.

Two giant arcs formed by a galaxy-cluster of velocity dispersion $1100 kms^{-1}$, core radius of 100 kpc and eccentricity of 0.85 situated at redshift of 0.23 are shown here. The background object (S1) at redshift of 0.85 is imaged to a linear structure (I1) consisting of a bright head which is a image of the entire source and a tail consisting of two merging images of part of the source. The central thin line-like shape is the critical curve for marginal lensing and the inner thick dashed curve is the corresponding caustic in the image plane. The image mergers occur at 1A and 1B. The object (S2) at redshift 1.36 is imaged into a feature (I2) consisting of three distinct images separated by brakes 2A and 2B. The respective critical curve and caustic are the lip-shaped thin dashed curve and the outer thick dashed curve.

More than one models are available for most of the above systems. However, considering the various available constraints I believe that in majority of these cases one single very low ($2 \times 10^{10} M_\odot$) to moderately high ($10^{12} M_\odot$) mass galaxy with large projected eccentricity (> 0.9) along with a moderately massive nuclear region ($\sim 10^{9} M_\odot$) best models the system reasonably well. Numerical stability of models at milliarcsecond scales due to inhomogeneities in the lens (Narasimha & Chitre, 1991) do not allow verification of the models to high accuracy unless extended features at low flux levels are imaged to high accuracy. Within this limitation, the models of these systems suggest that spirals and possibly LMC type of gas rich irregulars have the correct mass and core radius to produce the angular separation observed in these systems and the morphology of the system indicates that the lens mass is by and large distributed with high eccentricity. Edge on spirals naturally have the projected surface mass density as well as ellipticity to form these type of systems. When the eccentricity is near unity, the extended lip or diamond shaped critical curve has large area and these galaxies can form eye-catching features through their lens action on milliarcsecond jets or knots of background sources. Deep survey of radio sources at subarcsecond resolution should reveal more such systems, where the morphology of the images could be the best indicators of the lens (Narasimha, 1994a).

All these radio sources exhibit a few percent level of polarization. Polarization mappings provide useful constraints in modeling; more important, *correlation between images of changes in polarization characteristics at good enough, spatial resolution is a reliable technique to measure time-delay between them* (Nair *et al*, 1993, Wilkinson *et al*, 1994).

All of these lenses should exhibit unusually large Faraday rotation, which appears to be the case for at least 1830 − 211 and 0218 + 357. The differential Faraday rotation as a function of the lens redshift could be a probe of the evolution of magnetic field in galaxies at high redshifts. This should help settle the origin of cosmic magnetic field and also at what scales are they ordered.

6. Large Scale Geometry of the Universe from Cluster Lenses

Compact but rich clusters of galaxies at redshift upwards of ~ 0.2 have surface mass density comparable to the critical value for multiple image formation of sources at large distance. Consequently, many of the X-ray selected clusters have exhibited arc-like faint structures which are distorted background galaxies. If a galaxy is located nearly along the centre of the cluster and the D_{eff} for the system is suitable for marginal lensing or slightly above, the galaxy is magnified by ~ 50 times along the major axis of the cluster and it appears like a giant arc. For slightly higher redshift, it can form multiple images, all merging. (As an example, the linear features displayed in Fig.2 are possible for a range of source redshifts between 0.8 and 1.4 for the lens parameters specified there. However, observationally, depending on the source size the images could be distinct or a single linear feature).

The marginal lens properties are not sensitive to the details of the lens mass distribution, though the exact figures like velocity dispersion will depend on the details. However, the only important differences to be noted are whether the lens action is dominant by the smooth mass distribution in the cluster or it is governed by compact objects like cD galaxies. If one or more cD galaxy like structures having mass upwards of $10^{13} M_\odot$ and core radius less than 10 kpc provide the dominant contribution to lensing, the marginal critical curve will not be an extended lip; it will be more like beak-to-beak structure (Kaissola, Kovner & Blandford, 1992). However, if the smooth large scale mass distribution (> 100 *kpc*) determines the marginal lens caustics, the lip will be important for a range of redshifts as noted earlier (Narasimha, 1993; Narasimha & Chitre, 1993a,b). In realistic cluster lenses, the relative contribution of the two components is still not known, the main uncertainty being in deciding whether a cluster is relaxed.

If the smooth mass dominates the lens potential near marginal lensing, clusters offer a method to determine the large scale geometry of the Universe. The minimum mass density needed for multiple image formation by an elliptical mass is

$$\Sigma = \frac{1 + \cos(\beta)}{2} \Sigma_{cr} \qquad (18)$$

(Subramanian & Cowling, 1986), where $\sin(\beta)$ is the eccentricity of the mass distribution. Let us assume that for a more general but smooth mass dominated distribution, κ_c be the convergence (Eq.(14) at marginal value. The critical curve evolution through the change of source redshift was shown in Fig.2 as well as qualitatively described in section 5. If any background source is located within the critical curve close to the centre of the lens and near the marginal redshift, the multiple images will exhibit either breaks (2A and 2B) or mergers at similar positions or a combination of break (1A) and merger (1B). The separation between these points (2A to 2B or 1A to 1B) will be a measure of the image separation for a point source. This separation will vary as $\sqrt{\kappa_s - \kappa_c}$. For a given cluster but varying source distance,

$$\kappa_s \propto \frac{D_{LS}}{D_S}$$

Consequently, within the core radius of a rich galaxy cluster and with in a range of ~ 0.3 redshift from the marginal value, if there are three linear features, high resolution observations to determine the points of breaks or mergers will enable us determine the quantity $q(z_1, z_2, z_3; z_L)$ which is defined by

$$q = \frac{\kappa_2 - \kappa_1}{\kappa_3 - \kappa_1} = \frac{y_2^2 - y_1^2}{y_3^2 - y_1^2} \tag{19}$$

where y_i denote the distance between points of mergers or breaks for the ith image. With in our idealisations, the above ratio is a function of the redshifs of the sources and the lens as well as the cosmological parameter q_o. But it is independent of the cluster mass or radius or Hubble constant.

In realistic clusters, the arcs will not be positioned optimally. But at faint levels, the number of arcs in the field of the cluster close to its centre can be more than three and additional constraints like the ellipticity of mass distribution also may be available. Consequently, statistical analysis of $q(z_1, z_2, z_3, z_L)$ for large number of clusters dominated by smooth mass distribution should provide good constraints on the large scale geometry of the Universe, *independent of the Hubble constant.*

7. Weak Lensing and Lens Inversion

It has been pointed out in the previous section that a rich cluster at moderate redshift can produce multiple images or large magnification of distant sources along the line of sight to its centre. These clusters will distort images of background galaxies located at a few arcminutes of their field (*weak lens*). A deep enough ccd image of such a field can exhibit upto a few thousand slightly distorted galaxies, though usually the number is a few tens. If the cluster is not a high redshift and is not very rich, then most of the distorted background galaxies can be treated as at constant D_{eff}. Using the theory developed earlier, we can analyse the distortion of the images to regain the mass distribution of the lens (Kaiser & Squire, 1993).

The magnification matrix $[A]^{-1}$ (Eq.(17)) can be expressed as

$$[A] = \begin{bmatrix} 1-\kappa-\gamma_1 & -\gamma_2 \\ -\gamma_2 & 1-\kappa+\gamma_1 \end{bmatrix} \tag{20}$$

where $\gamma = \begin{pmatrix} \gamma_1 \\ \gamma_2 \end{pmatrix}$ is the shear due to the lens, which represents both distortion of the image and a rotation. The lens shear γ is related to the observed eccentricity of a circular source by

$$\frac{\varepsilon^2}{[1-\varepsilon^2/2]} \sim \frac{4(1-\kappa)\gamma}{(1-\kappa)^2+\gamma^2} \tag{21}$$

which can be derived by taking the components of a circular source and computing its magnification by using Eq.(20).

If we the complex formalism and express the vectors γ and angles in the image plane x as complex numbers, the lens equation can be used to derive the following relation:

$$\gamma = \frac{1}{\pi} \int d^2 x' k(x')\, K(x, x')$$

Here the Kernel is defined by

$$K(x, x') = \frac{1}{(x^* - x'^*)^2}$$

where * denotes complex conjugation. This equation can be inverted to find the converge κ to be

$$\kappa(x) = \frac{1}{\pi} \int d^2 x' \, Real\, [K^*(x, x')\, \gamma(x')] \tag{22}$$

If the number of distorted galaxies in the cluster field is large, at every position where a background galaxy is seen we can compute the eccentricity of the image and hence a relation between κ and γ which can formally be used to find γ. Using the information on the whole plane (i.e., all the positions where distorted images are seen) Eq. (22) can be used to infer κ which is proportional to the surface mass density of the cluster.

1. Boundary effects, the cluster centre could be a strong lens region and at outer parts of the cluster we might be tracing only the distortion due to the total mass.

2. The galaxies have intrinsic ellipticity which is much larger than the distortion due to weak lensing.

3. The effective distance for lenses at different distances can vary substantially. However, with improved observations, using strong lensing in the interior and weak lensing at the outer parts, it should be possible to get reasonable results relating to mass distribution in clusters.

8. Summary

Gravitational Lens promises to be a valuable probe of structures at mass scale ranging from less than 10^{-4} to greater than $10^{16} M_\odot$ and distance scales of kiloparsecs to the size of our Universe. The main hope in this field is because it does not depend upon the conventional methods of determination of mass or distances. The point lens, though simple, adequately describes microlensing and can be used to probe the structure of the Galactic Disk and Halo and possibly the centre too through microlensing. Gravitational microlensing is expected to provide a reliable technique to estimate the mass distribution of compact lenses as well as the size of the background variable sources. The Macho experiments have strongly indicated that the mass density of compact objects in the Galactic Disk is substantially higher than that implied by the conventional models. However, no firm limits are available for the Halo mass distribution. Lensing of VLB scale structures could show direct evidence for possible existence of blackholes in lens galaxies. Long term monitoring of AGNs lensed by galaxies en route provide a means to estimate the source size and brightness distribution of the central emission region of the AGN by analysing the microlens light curves.

The present results from more than one decade of observation and analysis of galactic scale lenses have not shown us results as unambiguous as were expected when the systems were discovered. However, cleaner and promising arcsecond and subarcsecond systems, even though observationally more challenging, could eventually provide the key to cosmic magnetic fields, galaxy formation and the cosmic length scale.

The cluster lenses have opened new avenues to study lensed galaxies at high redshift. Rotation curves of background spirals, typically magnified by 50 times by the cluster lens, can be studied and the evolution of the relation between rotation velocity and surface brightness for spirals as function of redshift should now be possible. The cluster mass distribution can be probed by weak lens and by the analysis of giant arcs. Possibly, multiple giant arcs observed in the field of rich clusters could provide a reliable method to estimate the large scale geometry of the Universe if the lens potential near the arcs is determined mainly by the smooth mass distribution in the cluster.

Acknowledgement

It is a pleasure to thank S.M. Chitre for valuable suggestions.

References

1. Alcock C. *et al*, 1993, *Nature* **365**, 621.
2. Alcock C. *et al*, 1994, *ApJ* (to appear).
3. Aubourg E. *et al*, 1993, *Nature* **365**, 623.
4. Bourassa R.R., Kantowaki R., 1975, *ApJ* **195**, 13.
5. Calder, R. it *et al*, 1993, in *Gravitational Lenses in the Universe*, proceedings, Liege, eds. J. Surdej *et al*, p325.
6. Henstock, D.R. *et al*, 1993, in *Gravitational Lenses in the Universe*, Proceedings, Liege, eds. J. Surdeg *et al*, p325.
7. Irwin, M.J. *et al*., 1989, *AJ* **98** 1989.
8. Jarosynski, M., Wambsganss, J. & Paczynski, B., 1992, *ApJ* **396**, L65.
9. Kaiser, N. & Squire, G., 1993, *ApJ* 404441.
10. Kaissola, A., Kovner, I. & Blandford, R.D., 1992, *ApJ* **396**, 10.
11. Kovner, I., 1987 *ApJ* **321**, 687.
12. Lynds, R. & Petrosian, V., 1986, BAAS **18**, 1014.
13. Mellier Y., Fort B., Kneib J.-P., 1993, *ApJ* **407**, 33.
14. Mellier Y., Dentel-Fort M., Fort B., Bonnet H., 1994, to appear in A & A.
15. Nair, S., 1993, *Thesis, University of Bombay*
16. Nair, S., Narasimha, D. & Rao, A.P., 1993, *ApJ* **407**, 46.
17. Narasimha, D., 1993, *Current Science* **64**, 725.
18. Narasimha, D., 1994a, in *ICNAPP* (in press) *ed. R. Cowsik*
19. Narasimha, D., 1994b, in *Bull Astr Soc India* (in press) *Proceedings of the ASI meeting*, ed. V.K.Kapahi.
20. Narasimha D., & Chitre S.M., 1988, ApJ **332**, 75.
21. Narasimha D., Chitre S.M., 1991, in *Gravitational Lenses*, eds., R. Kayser, T. Schramm, L. Nieser p378.
22. Narasimha, D., Chitre S.M., 1993a, J. A & A **14**, 121.
23. Narasimha D., Chitre S.M., 1993b, A & A **280**, 57.
24. Narasimha D., & Patnailk, A.R., 1993, Gravitational Lenses in the Universe, proceedings, Liege, *eds. J. Surdej et al,* p295.
25. Narasimha D., Subramanian, K, Chitre, S.M., 1982, MNRAS **201**, 941.
26. Narasimha D., Subramanian, K, Chitre, S.M., 1984, *MNRAS* **210**, 79.
27. Narasimha D., Subramanian, K, Chitre, S.M., 1986, *Nature* **321**, 45.

28. Paczynski, B., 1986, *ApJ* **304**, 1.
29. Paczynski, B. *et al*, 1994, *ApJ lett* **435**, L113.
30. Patnaik, A.R., 1993, in *Gravitational Lenses in the Universe*, proceedings, Liege, *eds. J. Surdej et al*, p311.
31. Patnaik A.R. *et al*, 1994, submitted to *MNRAS*.
32. Rauch, K.P. & Blandford, R.D., 1991, *ApJ* **381**, L39.
33. Saslaw, Narasimha & Chitre, 1985, *ApJ* **292**, 348.
34. Schild R.E. & Weekes, T., 1984, *ApJ* **277**, 481.
35. Soucail G. Fort B., 1991, A & A **243**, 23.
36. Soucail G. *et al*, 1987, A & A **172**, L14.
37. Subrahmanyan, R., Narasimha, D., Rao, A.P & Swarup, G., 1990, *MNRAS* **246**, 263.
38. Subramanian, K. & Cowling, S.A., 1986, *MNRAS* **210**, 333.
39. Udalski A. *et al*, 1993, *Acta Astr.* **43**, 289.
40. Udalski A. *et al*, 1994, *Acta Astr.* **44**, 165.
41. Walsh D., Carswell R.F., Weymann R.J., 1979, *Nature* **279**, 381.
42. Wilkinson, P. *et al*, 1994, in IAU *General Assembly.*

Alternatives to Dark Matter and Some Implications For the Early Universe

C. Sivaram
Indian Institute of Astrophysics
Bangalore 560 034, India

Abstract

Some consequences of MOND (Modification of Newtonian dynamics) proposed as an alternative hypothesis to dark matter for various astrophysical situations is discussed. The ubiquitous occurrence of the fundamental acceleration invoked is pointed out, for various physical situations in cosmology and physics. A possible connection with the cosmological constant and a vacuum dominated $\Omega = 1$ universe is discussed. Galaxy formation is considered in the context of MOND. Other alternative suggestions are critically discussed including Weyl gravity on a large scale, general relativity with torsion, gravimagnetic forces, etc. Implications of all these ideas for the early universe and constraints therefrom are also discussed.

1. Introduction

As often mentioned, the innumerable postulated candidates for dark matter cover a range of about eighty orders in mass! This is from the $\sim 10^{-40}g$. axion to $\sim 10^6 M_\odot$ black holes. Even the non-baryonic particle physics candidates range from micro eV axions to cento PeV CHAMPS! There are several ongoing searches with many clever experimental techniques to look for such particles. Some of the experiments have already been going on for years. So far there is no confirming evidence. Several possible candidates have been ruled out while on others only upper limits have been put! For instance in searching for super heavy isotopes using concentrated heavy water present limits on TeV champs are $< 10^{-30}$ (per baryon). A main motivation for non-baryonic DM is inflation (requiring $\Omega = 1$) which has emerged a favourite paradigm to avoid the finetuning FLATNESS problem. However, presence of vast amounts of dark matter (DM) would also pose equally serious "fine-tuning" problems. For instance why should $\rho_{dark} \approx 10^2 \rho_{visible}$, i.e. $\rho_{Non\text{-}bary} \approx 10^2 \rho_{Bary}$ rather than 10^{-8} or 10^{20}? This aspect of the problem is often ignored and presumably has to do with α_{GUTS}, i.e. the GUT scale.

We may have a possible analogy with supernovae Type II (SN II). In a typical SN II, the binding energy of the neutron star formed, i.e. $\sim 3\times10^{53}$ ergs is radiated chiefly in form of neutrinos and this radiation is $> 10^2$ times than in other forms, i.e. the 'DM emission' in this case is typically $> 10^2$ times than in visible radiation. Here we understand the basic reason for this, i.e. the weak interaction processes producing neutrinos dominate under such conditions! Similar thing can happen in the early universe if various types of defects form and decay predominantly into non-baryonic matter. Recently there have been revival of the suggestion that a residual cosmological constant or $\wedge - term$ can dominate cosmological dynamics. Even though $\wedge_{now}/\wedge_{Planck} \sim 10^{-122}$, even this can contribute as much as $\Omega = 0.8$ (i.e. $0.8\,\rho_c$) so that $\wedge_{everything} + \rho_{vacuum(\wedge)} = \rho_c$ satisfying $\Omega = 1$. Even this hypothesis is definitely not ruled out! It predicts more lensing events! Variation of angular diameter with redshift can help test this (for $\wedge = 0$, maximum at Z = 1.25 for $\Omega = 1$), like recent work of Kellerman. Another advantage of a $\wedge -$ term is that it can increase age of universe even with $H_0 \approx 100 km/s/MPC$.

Faced with so many alternatives for dark matter, MOND was postulated, keeping in mind the Vulcan problem in the solar system when Einstein's general relativity (modifying Newtonian law) removed the need for dark matter (i.e. Vulcan in this case!) to explain the anomaly in Mercury's orbit.

2. MOND and Some Consequences

'Modification of Newtonian dynamics' or MOND was initially proposed by Milgrom (1983) as an alternative hypothesis to account for the flat rotation curves of spiral galaxies without the need to invoke halo dark matter with mass progressively increasing with radius. It has had phenomenological success in explaining a wide variety of observations involving galaxies and galactic clusters including known empirical relations involving luminosity-velocity dispersion or luminosity-rotational speed (Milgrom and Sanders 1993).

The theory however requires the ad hoc introduction of a fundamental (critical) acceleration a_0, below which (a_0 is empirically about 10^{-8} cm sec^{-2}) Newtonian dynamics is modified to an effective 1/r force law dependence (rather than the usual $1/r^2$ one).

Thus we have

$$a = GM/r^2 \qquad a >> a_0$$

and

$$a = \frac{(GMa_0)^{1/2}}{r} \qquad (a << a_0) \tag{1}$$

where a_0 is introduced as a new fundamental constant having a value $\approx 10^{-8} cm sec^{-2}$.

Thus this give rise to a constant rotation speed ($V_c^2 \approx ar$) at the periphery of large spiral galaxies of :

$$V_C \approx (GMa_0)^{1/4} \tag{2}$$

Unlike some of the earlier modifications of gravity law (like for instance $a = GM/r^2 \ (1 + \frac{r}{r_0})$, where for $r >> r_0$ ($r_0 \approx 20 Kpc$) giving a constant circular speed $V_c = (GM/r_0)^{1/2}$,) the relation for V_C given by eq.(2) satisfies the infra red Tully-Fisher Law:

$$V_C \quad \propto \quad L^{1/4} \tag{3}$$

(L is the total luminosity of the galaxy).

The modification in the gravity law occurs whenever the acceleration drops below a_0 (i.e. low accelerations) rather than at large distances $r > r_0$.

As stated above this is claimed to explain a whole range of data from clusters to individual galaxies. Only major input is the assumption of $a_0 \approx 10^{-8} cm sec^{-2}$ albeit an arbitrary feature, although Milgram and Sanders (1993) note that it is significant that a_0 turns out to be close to the cH_0, where c is the velocity of light and H_0 is the Hubble constant suggesting a cosmological link. A lot is claimed to be explained with this one new constant. We point out the ubiquitous occurrence of a_0 in various physical situations.

A minimal a_0 can arise from cosmological considerations combined with the quantum uncertainty principle (Sivaram (1982, 83, 85)). For a typical elementary particle (say a hadron) of mass m, the gravitational self energy, i.e.

$$E_G \sim \frac{Gm^2}{(\hbar/mc)} \sim \frac{Gm^3 c}{\hbar} \approx \hbar H_0 \tag{4}$$

in order to be measurable over a Hubble time must be ($\approx \hbar H_0$) thus giving Weinberg's empirical relation

$$m_\pi \approx \left(\frac{\hbar^2 H_0}{Gc} \right)^{1/3} \tag{5}$$

(m_π is the pion mass).

This implies a 'surface gravity' (acceleration) for a hadron of:

$$a \sim Gm/r^2 \sim \frac{Gm^3 c^2}{\hbar^2} \sim \frac{Gm^3 c}{\hbar} \cdot \frac{c}{\hbar} \approx cH_0 \tag{6}$$

(from eq.(4)). Again surface gravity or acceleration due to gravity for an electron is: (self gravity) (mass m_e).

$$a_G\, c \approx \frac{Gm_e}{(e^2/2m_e c^2)^2} \approx \frac{4Gm_e^3 c^4}{e^4} \approx 10^{-8} cmsec^{-2} = a_0 \tag{7}$$

Also for a typical atomic nucleus (mass m_n) ($A \approx 150$) , surface gravity

$$\approx \frac{Gm_n}{r_n^2} \approx 10^{-8} cmsec^{-2} \approx a_0 \tag{8}$$

'a_0' of course has an ubiquitous occurrence over several large scale structures.

For a typical spiral galaxy ($M_{gal} \approx 10^{12} M_{\odot}$, $R \approx 30Kpc$)

$$a \approx \frac{GM_{gal}}{R^2} \approx \frac{6.6.10^{-8}.2.10^{45}}{(10^{23})^2} \approx 10^{-8} cmsec^{-2} = a_0 \tag{9}$$

Also for a typical cluster $M_c \approx 10^{16} M_{\odot}$ $R \approx 3MPC, a \approx 10^{-8} cmsec^{-2} = a_0$.

Again for globular cluster $R \sim 100pc$ $M \sim 10^6 M_{\odot}$, $a \approx a_0$. An underlying reason for this ubiquitous occurrence of a_0 is essentially that for a system of mass density ρ and spatial extent R, the gravitational self acceleration is:

$$a \approx G \rho R \tag{10}$$

and if $\rho R =$ constant for a given set of objects or structures or systems then $a_0 \approx 10^{-8} cmsec^{-2}$. Also for the universe as a whole ($\rho \approx 10^{-29} g.cm^{-3}$, $R \approx 10^{28} cms$ so $\rho R \sim 0.1 gcm^{-2}$) implying a deceleration from the Robertson-Walker equation of:

$$\ddot{R} \approx 4\pi G(\rho + P) R \approx 10^{-8} cmsec^{-2} = a_0 \tag{11}$$

This brings us to the notion of surface density. It is to be noted that the surface density

$$\sum = \frac{M}{R^2} = \frac{a_0}{G} = \text{constant} \tag{12}$$

for all of the above systems.

MOND predicts universality of Σ, satisfying the so called Freeman-Fish empirical law on surface brightness of ellipticals and spirals.

Thus we have:

$$\sum_{galaxy} \approx 0.15 gcm^{-2}$$

$$\sum_{galaxycluster} \approx 0.12 gcm^{-2}$$

Surprisingly

$$\sum_{electron} \approx 0.1 gcm^{-2}, \sum_{proton} \approx 0.2 gcm^{-2}$$

Also from eq. (6) for a typical hadron.

$$\sum_{hadron} = a_0/G = \sum_{gal} = \sum_{universe} = \sum_{cluster} = \sum_{globularcluster} \text{ etc.} \tag{13}$$

Equations (12) and (13) also imply the universal micro-macro systems relation: (as $1/r^2 \approx$ curvature):

$$\text{Energy} \times \text{curvature} = \frac{c^2}{G} \cdot a_0 = \text{universal constant}$$

It is to be noted that c^2/G is the superstring tension (i.e. fundamental superstrings formed at the Planck scale have a tension $\sim c^2/G$) (Sivaram, 87, 90, 93).

Energy $\times$ curvature = Tension $\times\, a_0$ = same for all scales in the universe (eqs.(12) and (13)?).

To test whether there is really deviation from Newtonian dynamics look for (or create.) systems where surface gravity or acceleration becomes $\sim a_0$ or less. Let us recall that MOND implies modification at low accelerations and not large distances.

For instance a spacecraft orbiting sun at $\sim 10^4 A.U. \approx 10^{17} cms$ has $\omega^2 R \approx a_0 \approx 10^{-8} cmsec^{-2}$, with orbital velocity $\sim 0.3 km/sec$. The pioneer and Voyager spacecraft are at present distances less than 10^2 A.U. With present technology a minimal energy trip for a spacecraft to reach 10^4 A.U would take $\sim 10^5$ years. If launched at velocity $\approx 10^3 km/sec$, the time taken to reach orbit at 10^4 A.U. is ~ 50 years.

A probe launched $\sim 10^3\, km$ away from a 10 Km asteroid is subject to acceleration $\sim 10^{-8} cmsec^{-2}$. In principle these launches are possible to test for deviations from Keplerian dynamics at low accelerations. If there is a OORT cloud of comets at 5 x 10^4 A.U. or 10^5 A.U. then the orbital accelerations $a < a_0$. This would thus involve deviations from Newtonian dynamics as envisaged by MOND on solar system scales. (Sivaram 1993). In an orbiting zero gravity space laboratory a 1 kg.mass and a one gm. mass separated by about a metre would experience a mutual gravitational acceleration $a < a_0$. In the microgravity environment deviations in free fall time can be used to look for such effects. Two particles of radii r, equal masses m and separation 'd' would free-fall under their gravitational attraction in a time.

$$t_{ff} = \frac{1}{3}(d^3/GM)^{1/2} - 2/3\,(2r^3/GM)^{1/2}$$

Also suitable laser beams could be used on board the space station to give minute accelerations to test objects. These would be explored elsewhere.

Again accelerations $< a_0$ can also occur in star forming regions over distances of a few parsecs. Consider a collapsing gas cloud (with temperature T, density ρ and total energy E_T).

$$E_T = E_{thermal} + E_{grav} = M\,[3R_g\,T - \frac{1}{2}\,G(\frac{4}{3}\,\pi\,\rho)^{1/3}\,M^{2/3}]$$

For it to be gravitationally bound, $E_T < 0$; so that for $T \approx 10\,°K$, $\rho \sim 1$ atom cm^{-3}, the radius R and mass M of gas sufficient for gravitational binding is $R \approx 30pc$ and $M \approx 10^3 M_\odot$. Such a cloud has surface gravity $a_0 \approx 10^{-8} cmsec^{-2}$ at $R \approx 5pc$, which implies surface gravity $< a_0$ for such a cloud. There could thus be consequences of MOND in such situations.

There are several other astrophysical situations where a_0 arises naturally. To give a few more instances:

1. The drag force on a proton moving through the intergalactic medium gives an acceleration:

$$a \approx \frac{\rho \cdot cross - section \cdot V^2}{m_P} \times constant(\sim 1)$$

$$\approx 10^{-29} \times 10^{21} \times 10^{-24}/10^{-24} \sim 10^{-8} cm/sec^2 \approx a_0$$

(assuming $\rho \approx 10^{-29} gcm^{-3}$, $V \approx c$ and geometric $C.S. \approx 10^{-24} cm^2$.)

2. The acceleration due to radiation drag on an electron due to cosmic microwave background is:

$$a \approx \sigma_T c^2 \rho_{rad}/\alpha\, m_e$$

where $\rho_{rad} \approx aT^4_{rad}$ (For $T_{rad} \approx 2.7°K$, $\rho_{rad} \approx 4.5 \times 10^{-13} ergcm^{-3}$
σ_T is the Thompson C.S. a_0 turns out to be $\approx a_0 \approx cH_0 \approx 10^{-8} cmsec^{-2}$

3. The smallest acceleration which can cause barely detectable gravitational radiation of power $\sim \hbar H_0^2$ (as given by the uncertainty principle is also a_0).

Thus we have:

$$P_G \approx \frac{G}{c^5} M^2 \omega^6 R^4 \approx P_{G min} \approx \hbar H_0^2$$

Using $\omega^2 R = a$, $\omega^2 R^3 = GM$, $\omega^6 R^3 = a^3$, and using for M, $M_{pl} = (\hbar c/G)^{1/2}$ as the smallest particle mass for which gravity is dominant, so that $GM_{pl}^2 \approx \hbar c$, we finally have: $\hbar/c^4 \cdot a^3 R \approx \hbar H_0^2$, and choosing maximum value $R \approx c/H_0$, we get $a_{\min} \approx cH_0 \approx a_0$.

3. MOND and Galaxy Formation

There are also consequences of MOND for galaxy formation. Standard parameters like the Jeans length get modified. The pressure is now given by an expression like:

$$P \approx \frac{(GMa_0)^{1/2} M}{r^3} \tag{14}$$

This can be shown to give rise to a modified Jean's length of: (Sivaram 1993b)

$$L'_J \approx \left(\frac{c_s^4}{Ga_0\rho}\right)^{1/3} \tag{15}$$

where c_s is the sonic speed in the medium. (compared to the usual $L_J \approx c_s/(G\rho)^{1/2}$). The corresponding modified Jeans mass is:

$$M'_J \approx 4\pi/3 \left(\frac{c_s^4 \rho^2}{Ga_0}\right)^{1/3} \tag{16}$$

This gives around recombination, $L'_J \approx 10^{25}\, cm$ and $M'_J \approx 10^{19} M_\odot$, implying somewhat larger structures than in the usual scenario. Again the growth of density fluctuations in an expanding universe can be studied with the MOND force law. This gives (Sivaram 1993b) the growth rate equation for a density fluctuation $\rho^1 = \rho(1+\delta)$; which translates into:

$$\ddot{\delta} + 2\dot{\delta}\frac{\dot{R}}{R} = \delta\,(Ga_0)^{1/2}\,\rho^{1/2} R^{-1/2} \tag{17}$$

$$\ddot{\delta} + 4\frac{\dot{\delta}}{3t} = \frac{const \cdot \delta}{t^{4/3}} \quad \ldots. \tag{18}$$

Solution is still given by approximate power laws (of type $\delta \approx At^n + Bt^{-n}$), which implies there is not much improvement in rate of growth of fluctuations as in standard scenario without vast amounts of cold dark matter, etc. This could be a major drawback: although flat rotation curves and cluster dynamics can be explained by MOND it does not seem to improve the situation as far as formation of large scale structures is concerned.

4. MOND and the Cosmological Constant

Recently there have been revivals of the suggestion that a residual cosmological constant dominates dynamics. As noted above, MOND can probably account for galactic rotation curves and cluster dynamics without the need to invoke dark matter. However, a closed $\Omega = 1$ universe with $\rho = \rho_c$ as favoured by the inflation models of the early universe (to account for the flatness, horizon and other problems, would still need vast amounts of dark matter on cosmological scales (the baryonic matter is constrained from various arguments to be $< 0.1\rho_c$). MOND does not directly settle this question. As mentioned in the introduction several ongoing searches over the past couple of years to directly detect dark matter have not been successful and besides there are now far too many possible candidates! This is one of the reasons for reviving the suggestion that a residual $\wedge - term$ can contribute to as much as $0.8\rho_c \; or \; \Omega = 0.8$ or more (Yoshii, 1993a, b) so that:

$$\rho_{everything} + \rho_{vacuum}(\lambda) = \rho_c \tag{19}$$

The $\wedge - term$, as is well known, is associated (see the discussion in Sivaram (1986a,b) with a vacuum energy density given by:

$$\rho_{vac} \approx \frac{\wedge c^4}{8\pi G} \tag{20}$$

Thus a $\wedge - term$ (units of curvature cm^{-2}) of magnitude $\wedge \approx 10^{-57}$ would imply a $I\rho_{vac} \approx 10^{-29} gcm^{-3} \approx \rho_c$.

Such a background cosmological constant term apart from making up the critical density can also provide a basis for the existence of the fundamental acceleration a_0 assumed ad hoc in MOND.

The force acting on any particle of mass m in the universe because of the existence of a $\wedge - term$ is given through de-sitter metric as:

$$F \approx \frac{c^2 \wedge r}{m} \tag{21}$$

so that one gets an acceleration (which is mass independent, as $\wedge$ acts universally and does not violate the equivalence principle) of:

$$a \approx c^2 \wedge r \tag{22}$$

For $\wedge \approx 10^{-57}\, cm^{-2}$, $r \approx R_H \approx 10^{28} cm$ (Hubble radius)

$$a \approx 10^{-8}\ cms^{-2} \approx a_0$$

This suggests that the origin of a_0 on a cosmic scale might be traced to a $\wedge - term$ of this magnitude which contributes to most of the critical density. A physical fundamental basis for the vacuum energy dominating in a $\Omega = 1$, $\rho = \rho_c$ universe was explored in Sivaram (1986c).

Again eqs. (21) and (22) would imply a force or acceleration on large scales $r \sim 1/\sqrt{\wedge}$ or $\wedge \sim 1/r^2$ which is proportional to $1/r$ rather than $1/r^2$ as in newtonian dynamics. MOND also implies a $1/r$ force law. These consequences of the $\wedge - term$ are tantamount to a modification of the Poison equation as:

$$\nabla^2 \phi + \wedge c^2 = 4\pi G \rho$$

and not

$$\Delta^2 \phi + \wedge \phi = 4\pi G \rho$$

as wrongly written by most authors including the earliest mention of it by Einstein. The modification can easily be accommodated by using a variational principle.

Thus MOND supplemented by a cosmological term can do away with the need for vast amounts of dark matter on cosmological scales also, not just on galactic and cluster scales. This term may also provide a basis for a_0 and explain another curious coincidence noted in Sivaram (1994a,b): i.e. the gravitational self energy density of a wide range of systems, i.e. an elementary particle, a galaxy or a galactic cluster are all the same and equal to the critical cosmological matter density i.e.

$$\rho_G \approx \rho_c \approx \frac{3 H_0^2}{8\pi G} \tag{23}$$

we note that the gravitational self energy density of a system is:

$$\rho_G \approx \frac{GM^2}{R^4} \tag{24}$$

which can be expressed in terms of the surface gravity or acceleration ($a = GM/R^2$) as:

$$\rho_G \approx \frac{a^2}{G} \tag{25}$$

and since $a \approx a_o \approx cH_0 =$ constant, universally for each of the systems in eqs. (13) it follows from eqs. (13) and (25) that the gravitational self energy density is also surprisingly the same for an elementary particle or for a galaxy or supercluster and is given universally by:

$$\rho_G \approx a_0^2/G \approx 10^{-8} ergcm^{-3} \approx 10^{-29} gcm^{-3} \tag{26}$$

Eq.(26) gives the impression that ρ_G seems strikingly near the critical density ρ_c, i.e. $\rho_c \approx 3H_0^2/8\pi G \approx 10^{-29} gcm^{-3}$. That this is not surprising and indeed so, can be demonstrated by considering eq.(5) for a typical particle mass and considering its gravitational self-energy density as:

$$\rho_G \approx \frac{Gm^2}{2(\hbar/mc)\,(\hbar/mc)^3\,(4\pi/3)}$$

$$\approx \frac{3}{8\pi}\,\frac{Gm^6c^4}{\hbar^4} \tag{27}$$

Now substituting for m from eq.(5) into eq.(27) gives:

$$\rho_G \approx \frac{3\,H_0^2}{8\pi G}$$

which is indeed ρ_c. Also for the universe as a whole, i.e. for a $\Omega = 1$, $\rho = \rho_c$ closed universe, the gravitational self energy density again equals ρ_c, i.e. $(\rho_G)_{univ} \approx \rho_c$. Thus in this case the total energy is zero.

It may also be mentioned that a model with a dominating $\wedge$ *– term* has been shown to provide fair fits to faint number counts, redshifts distributions (Yoshii, 1993b) and has the additional cosmological advantage of enabling globular cluster ages to be less than the age of the universe!

5. Maxwellian Alternative to Galactic DM Problem

There have been suggestions that magnetic forces or even a gravo-inductive forces (eg. Fahr, 1990) which is the gravitational analogue of magnetic force B may act on matter in galactic halos giving rise to an effective ~ $1/r$ force law thus accounting for the flat rotation curves. As an example we shall consider the latter postulate. It was critically discussed in Sivaram (1993). A Lorentz type force relation was written for the mass m :

$$m \frac{d}{dt} \bar{v} = y_1 m E_g + y_2 m \frac{1}{c_g} (\bar{v} \times B_g)$$

B_g like a magnetic field is velocity dependent ($\alpha\ v$) . Being the 'radiation' part of the field it would fall off with distance only as 1 / r and not $1 / r^2$, so that balance with the centripetal force mv^2/r, would imply a constant v (independent of r) at distances larger than the outer boundary of the galactic core where this force is expected to dominate. To account for the flat rotation curves, y_2 must be $\sim 10^6 y_1$. However, it is well known that equations of motion in GR already contain gravi-magnetic forces. The geodesic equations can take a form similar to the Lorentz force with $E_g = \nabla V$, $B_g = \bar{v} \times \nabla \times A$ with $A_i = ch_{oi}$, $V = c^2 hoo$.

However unlike in electromagnetism, Lorentz type equations of motion are not independent of field equations. Same linearized metric tensor describes both forces and hence must have the same coupling constant.

$$\text{From } c = \frac{1}{\sqrt{\mu_g \varepsilon_g}}, \quad \varepsilon_g = \frac{1}{G}, \quad \mu_g = \frac{G}{c^2} = 7 \times 10^{-29}$$

which implies an extremely small gravity permeability coefficient. Thus this cannot give a ~ 10^6 enhancement.

Again composition dependent forces would give a coupling ~ 10^{-3} times weaker than E_g (Sivaram, 1993, 1990, Sivaram and Bertotti, 1991). $M = (A - Z)\, m_n + Z m_p, A m_p = M + (A - Z)\,(m_p - m_n)$.

The coupling has a factor $(m_n - m_p/m_n)^2 \sim 10^{-6}$ too a small to be useful even in early universe.

6. Weyl Gravity over Large Scales

Another solution (kazanas and Manheim, 1991) has involved postulating gravity theory quadratic curvature lagrangians (like in Weyl's theory) or in general non-linear curvature Lagrangians. This modification is considered valid over large scales > several kiloparsecs. For instance in the case of Weyl type theory with a R^2 action (R being the curvature scalar), the modified Poisson equation has a solution of the form

$$\nabla^{-4} m\delta^3(r) \sim MR,$$

i.e. Mass increasing linearly with R over large scales. This can be shown to account for flat rotation curves and account for cluster dynamics again without need for dark matter over large scales. Cosmologically this has the implication that only systems with zero *total* energy are allowed. Thus we naturally have a flat $\Omega = 1$ universe as an initial condition, i.e. the initial state would have been a fluctuation with zero total energy thus accounting for the flatness problem. Of course the scale invariance of the quadratic action would be broken at some scale say r_0 below which general relativity would be induced, giving the usual Newtonian $1/r^2$ force law at $r \ll r_0$. i.e. at smaller scales we would have a modified Keplerian law as:

$$\frac{V^2}{r} = \frac{d\phi}{dr} = \frac{GM}{r^2}[1 - k(1 + \frac{r}{r_0})\, e^{-r/r_0}]$$

(the modified Poisson equation is now of the form $\nabla^4\phi + \nabla^2\phi =$ constant, which gives the above solution). For $r \ll r_0$, we have the usual Keplerian $V \alpha\ 1/\sqrt{r}$. For $r \gg r_0$ $(r_0 \approx 10Kpc)$, V tends to a constant value and for a suitable choice of the constant K can be fitted with actual rotational curves with V tending, to a constant value. In general the above solution corresponds to actions of the form $R + R^2$. As is now well known, such actions have interesting implications for the early universe in the sense that they provide an alternative to the usual scalar-field dominated inflation, as first pointed out by Starobinsky (1980). Again the formal equivalence between gravity theory with *f(R)* actions and ordinary general relativity with massive self-interacting non-minimally coupled scalar fields has been well established (Barrow and Ottewil 1983, Sivaram 1979). Again theories with Lagrangians nonlinear in the curvature scalar can be reduced by appropriate transformations to a form similar to the Brans-Dicke theory which may enable us to have extended inflation in the early universe without the difficulties of introducing scalar fields (Sivaram, 1992) 1987). Briefly if we consider

$$L \sim f(R)$$

the field equations are of the form:

$$f'(R_{\alpha\beta} - \frac{1}{2}g_{\alpha\beta}R) + \frac{1}{2}g_{\alpha\beta}(Rf' - f) - \nabla_\alpha \nabla_\beta f' + g_{\alpha\beta} \Box f' = 0$$

Using the conformal transformations

$$g'_{\alpha\beta} = \lambda^2 g_{\alpha\beta}$$

(or $\lambda^2 = [f'(R)]^{1/2}$ so that $f'(R) = \lambda^4$) and defining $\phi = \frac{1}{2}\ln[f'(R)]$ the above equation becomes:

$$R'_{\alpha\beta} - \frac{1}{2} g'_{\alpha\beta}\, R = \frac{3}{2} \nabla_\alpha \phi \nabla_\beta \phi - \frac{3}{4} g'_{\alpha\beta}\, (\nabla_\mu \phi \nabla^\mu \phi) - \frac{1}{2} g'_{\alpha\beta} (f')^{-2} (Rf' - f)$$

where the effective gravitational constant in the above equations is defined in terms of the derivative of the function f of the curvature scalar as.

$$G_{eff} \sim (f'(R))^{-2}$$

and since R and ϕ are related above it is possible to express G_{eff} in the form:

$$G_{eff} \sim [f(\phi)]^{-1}$$

which is the form used in the extended and hyperextended inflationary scenarios. Thus this alternative solution to the dark matter problem may also provide a bonus in that it may give a natural basis for extended inflation in the early universe.

7. Gravitating Systems with Torsion Self-interaction

Another possible but highly tentative approach is to consider a minimal modification of general relativity, namely use of the asymmetric part of the connection i.e. the torsion:

$$Q^\mu_{\alpha\beta} = \Gamma^\mu_{[\alpha\,\beta]}$$

The torsion vector $\overline{Q}$ is related to the spin density σ of a matter distribution as:

$$\overline{Q} = 4\pi G\sigma/c^2$$

In the case of propagating torsion we can write: (Sivaram and de Sabbata, 1990, 1994).

$$\overline{Q} = \delta_\mu \phi$$

By analogy with electromagnetic E or B fields the energy density of the torsion field can be written:

$$Q^2 = const \cdot \delta_\mu \phi\, \delta^\mu \phi$$

The total torsional self energy is given by the volume integral. (Sivaram and de Andrade, 1993, 1994).

$$E_T = \int Q^2 \cdot dV$$

where: $Q^2 \ \alpha\ 4\pi\ G^2\ \sigma^2/c^4$ can be interpreted as a contribution to the energy density in the Poisson equation which thus gets modified to:

$$\nabla^2\phi \approx 4\pi G(\rho - \frac{G}{c^4}\sigma^2)$$

For the case of just gravitating torsional self energy, we have:

$$\nabla^2\ \phi_T \approx\ -4\pi G^2\sigma^2/c^4$$

which has the solution:

$$\phi_T \approx G^2\sigma^2\ r^2$$

If S is the total intrinsic spin of the source

$$\phi_T \sim G^2S^2\ /c^4r^4$$

Now S can be written $S \sim m\,\omega\, r^2$ (for a constant ω).

$$S^2\ \propto\ m^2\omega^2r^4$$

Hence the gravitational potential due to this additional interaction of spins is (V is the velocity)

$$Pot.\ \propto\ V^2\ \propto\ \frac{G^2S^2}{r^4}\ \propto\ \frac{G^2m^2\omega^2r^4}{r^4}\ \rightarrow\ constant$$

(as r cancels out!).

So this type of additional gravitational effect can in principle give rise to a velocity V that tends to become a constant independent of r (i.e. give rise to a flat rotation curve).

This is also borne out by an exact solution of the differential equation:

$$\nabla^2\phi \approx const \cdot \sigma^2 = const{\cdot}Q^2 = const \cdot (\delta_\mu\phi)^2$$

i.e. we have an equation of the type.

$$\delta^2\phi/\delta r^2 \approx const \cdot (\delta\phi/\delta r)^2$$

This can be seen to have a solution of the type:

$$\phi \approx const \cdot \log r$$

which gives rise to a force:

$$F \propto \frac{\delta\phi}{\delta r} \sim const \cdot \frac{1}{r},$$

which when balancing the centripetal force $\sim mv^2/r$, gives a constant velocity V independent of r. It turns out that modes of the torsion tensor interacting with matter correspond to a massive scalar field.

We can write down a Lagrangian for the torsion vector Q_μ containing Q_μ or its derivative as:

$$L_T \approx \alpha\, Q'_\mu Q^\mu + \beta(\delta_\mu Q^\mu)^2 + \gamma\, \delta_\mu Q_\nu\, \delta^\mu\, Q_\nu, \text{ etc.}$$

With $Q = \delta_\mu \phi$, we see that we have interaction terms similar to that invoked in extended inflation theories.

Thus once again it appears as if alternative solutions to the dark matter may also converge on to solutions of the problems of the early universe.

References

1. Milgrom.: 1983, Ap.J. **270**, 365.
2. Milgrom.: and Sanders, R.: 1993, Nature **362**, 25.
3. Sivaram, C.: 1982, Amer. J. Phys. **50**, 279.
4. Sivaram, C.: 1983, Amer. J. Phys. **51**, 277.
5. Sivaram, C.: 1986, Astrophys. Spc. Sci. **125**, 189.
6. Sivaram, C.: 1987, A.A. **37**, 201; Nature **327**, 108.
7. Sivaram, C.: 1993a, Astron. Astrophys. **275**, 37.
8. Sivaram, C.: 1994b, Astrophys. Spec. Sci. (in press).
9. Sivaram, C.: 1994c, (in preparation).
10. Sivaram, C.: 1993, Astrophys. Spc. Sci. **215**, 185.
11. Sivaram, C.: 1994, Astrophys. Spc. Sci. **219**, 135.
12. Sivaram, C. and de Andrade.: 1994, Gen. Rel. Grav. **26**, 615, and Class. Q. Grav. (in press).
13. Sivaram, C. and De Sabbata, V.: 1994, Spin and Torsion in Gravitation World Scientific Publisher. Singapore and London, 1994. (and referees therein).

14. Sivaram C.: 1994, Int. J. Theor. Phys. **33**, 2407.
15. Yoshii, Y.: 1993a, b, Ap. J. **403**, 522, **418**, L1.

Additional References

1. Sivaram, C.: 1986, Int. J. Theor. Phys. **25**, 825; **26**, 1425.
2. Sivaram, C.: Sivaram, C.: 1986, Astrophys. Spc. Sci. **127**, 133; **194**, 239.
3. Bertotti, B. and Sivaram, C.: 1991, Nuovocimento **106B**, 1299.
4. Fahr, H.: Astron. Astrophys. 1990, **236**, 88.

On the Final Fate of Gravitational Collapse

P.S. Joshi
Tata Institute of Fundamental Research
Homi Bhabha Road, Bombay 400 005, India

Abstract

We review here some recent developments in the field of gravitational collapse and the related cosmic censorship conjecture. The singularity problem in general relativity is discussed with reference to those forming in collapse situations.

The purpose of this review is to draw attention to several recent developments in the field of gravitational collapse and the related cosmic censorship conjecture. For further details, we refer to Ref. 1 and references therein.

When the stars such as Sun exhaust their nuclear fuel, they must collapse under the pull of gravity. For stars with masses of the order of a few solar masses, the end state would be a stable configuration such as a white dwarf or a neutron star. However, stars are known to have much larger masses, say fifty times the mass of the Sun. In such cases, if the star does not emit away almost all of its matter, there is no final stable configuration available and the collapse results into a spacetime singularity of infinite density. That such a singularity would always be hidden within a black hole when it forms in a gravitational collapse is the broad spirit of the cosmic censorship hypothesis of Penrose.

Thus, a particularly interesting question regarding the singularities occurring in gravitational collapse is whether these are necessarily covered by an event horizon and hence invisible to an out side observer at infinity. In the idealized case of the collapse of a homogeneous spherically symmetric dust ball[2], the space-time metric interior to the dust ball is given by the closed Friedmann metric, which is matched at the boundary by the exterior Schwarzschild metric. All the matter collapse to a space-time singularity at $r = 0$ which is completely covered by the event horizon at $r = 2m$. Such a singularity cannot emit any messages to an outside observer.

The statement that such a situation would persist even in a generic gravitational collapse, when departures from the assumptions such as homogeneity of the matter distribution, the dust form of matter etc. are allowed for is called the *cosmic censorship hypothesis*. A rigorous formulation and proof of the cosmic censorship hypothesis, ruling out the occurrence of naked singularities of gravitational collapse has been one of the most outstanding open problem in gravitation theory. Such a hypothesis lies at the very foundation of the theory and applications of the black holes.

The cosmic censorship was proposed in 1969 by Penrose, and when the attempts for obtaining the proof for the same started, the actual hope at the time was that it would be possible to establish a censorship theorem just similar to the singularity theorems being proved at that time in general relativity by Hawking, Penrose and Geroch[3]. The singularity theorems in general relativity establish the occurrence of space-time singularities in gravitational collapse and also in cosmological situations under a wide range of general conditions which have considerable physical plausibility. Such conditions include, typically, a suitable causality condition, the positivity of mass-energy density, the generic condition which implies that every non-spacelike trajectory must come across atleast once a non-zero stress-energy density, and a suitable convergence condition in the form of formation of trapped surfaces in the case of gravitational collapse. Such singularities then occur in the form of nonspacelike incomplete geodesics in the space-time. The proofs basically depend on two areas of study in order to infer the existence of such singularities. These are the topic of general global properties of a space-time obeying Einstein equations (which are used very weakly in the proofs of the singularity theorems) and the gravitational focusing of congruences of non-spacelike trajectories in a space-time.

While the singularity theorems established the existence of space-time singularities in the form of non-spacelike geodesic incompleteness, they were silent on the issue of their nature (e.g. Would the curvature invariants blow up near a space-time singularity? Will it be a physically significant singularity? Will it occur in future or past? etc.), and on whether these will be necessarily covered by the horizons. In fact, these theorems do not imply that the singularities forming in gravitational collapse must be necessarily hidden inside an event horizon, and hence invisible to an out side observer who is far away from the collapsing star. Such a situation requires a detailed analysis of the nature and structure of the space-time singularity which might occur under a specific collapse scenario and a determination is needed which would ensure whether the given singularity is physically meaningful or not in terms of the classification available for the physical significance of the singularity. Censorship proposed to prove in generality, or by assuming reasonable conditions such as the positivity of energy density etc. that singularities of gravitational collapse will always be covered by horizons of gravity, just as the Schwarzschild singularity is covered by the event horizon.

Desperate attempts for the proof of such a conjecture in general, in the form of a solid theorem, did not materialize in the next few years further to the censorship proposal, though people argued that there were indications that it should hold. Then, in 1974, Seifert et al[4] published a result showing that in the Tolman-Bondi solutions representing the inhomogeneous collapse of dust ($p = 0$), shells of matter will cross each other to produce momentary delta function like singularities (called 'shell-crossing' singularities), which were visible from the future null infinity, and hence naked.

The Tolman-Bondi space-times represent general spherically symmetric dust solutions of the Einstein equations. When we consider a dust collapse which is completely homogeneous, the solution is uniquely described by the Oppenheimer-Snyder space-time. However, as soon as some inhomogeneity is introduced (which would be a realistic case because no one expects a collapsing star to have a completely homogeneous distribution of density in space) in the dust distribution, things get quite complicated. Most interestingly, even for a completely general dust distribution, i.e. when ρ is a function of spatial coordinates, it is possible to solve the Einstein equations completely and this was done by Tolman in 1934 and Bondi looked at the problem again in 1948 in the context of collapsing dust clouds. These are known as the Tolman-Bondi solutions. Thus, this is a complete description of inhomogeneous dust collapse in terms of two free functions of the spatial coordinate r, which describe the degree of inhomogeneity of matter distribution at a given instance of time.

However, by the time Seifert et al reported the shell crossings of matter in Tolman-Bondi inhomogeneous collapse, a detailed classification of singularities, as regards their physical seriousness in terms of divergence of curvatures etc. was already available, and it turned out that these were gravitationally weak in the sense that the gravitational tidal forces did not diverge near such singularities. Further, as pointed out by Papapetrou and Hamoui[5] earlier in 1967, the space-time could be extended through such singularities (which were removable in this sense), and as pointed out by Hawking[6], the space-time metric could also be defined in the neighborhood of such a singularity in a distributional sense. Thus, such shell crossings by matter were not treated as serious threat to censorship, and as far as we know, the same attitude continues as of to-day, probably with some justification.

Of more serious nature were the 'shell-focusing' naked singularities which were pointed out by the numerical simulations of Eardley and Smarr[7]. This was again within the context of Tolman-Bondi space-times representing inhomogeneous dust collapse, and occurring at the center of the collapsing cloud at a sufficiently advanced stage of collapse. These singularities did not appear distributional like the shell crossings, and not removable in terms of singularity classification. An analytic treatment of this case was given by Christodoulou[8] in 1984 by working out radial null geodesics from the naked singularity. However, it was shown by Newman[9] in 1986 that actually Christodoulou considered a restricted subclass of Tolman-Bondi models (by imposing restrictions on the two free functions in these models describing the inhomogeneous matter distribution), and he also showed that these were again gravitationally weak in the sense that the gravitational tidal forces did not diverge. Specifically, one would want $R_{ij}V^iV^j$ to diverge as $1/k^2$, where k is the affine parameter along the trajectory, in order to classify the singularity as a strong curvature singularity in terms of the singularity classification; (see e.g. Ref. 3).

In fact, further to this development, it was conjectured by Newman and others that the cosmic censorship should be valid in the form that the naked singularities, even if they occur, must be gravitationally weak Around the same time (early 1980s), several attempts were made by different groups to establish the censorship hypothesis rigorously in one of the above forms, however, without success. Most of the times, these theorems contained rather restrictive conditions and anyway gave too little information.

Around the same time, a detailed investigation of the radiation collapse scenario was started (see e.g. Ref. 1), and it was pointed out that photon as well as particle trajectories would come out of the naked singularity forming at the center. The basic difference between this and the inhomogeneous dust collapse is that of equation of state, as here we have an equation of state describing a collapsing radiation. A detailed analysis of the structure of this naked singularity was completed by Dwivedi and Joshi[10], working out all the families of non-spacelike trajectories emerging from this naked singularity and also analyzing the curvature growth along the same. It turns out that there is a very powerful curvature growth along *all* these families, implying the divergence of gravitational tidal forces. Thus this turned out to be the most powerful curvature singularity in terms of curvature growth and as stated by the singularity classification. In fact, the curvatures grow near the singularity as fast and powerfully as in the case of the big bang or the Schwarzschild singularity. It was thus shown that the cosmic censorship cannot be true in the form that naked singularities can never be gravitationally strong.

Under the situation, we would argue that what is needed first is a detailed examination of collapse scenarios other than the exact homogeneous dust collapse case to examine the possibilities arising. Only this could provide insights into the issue of the final fate of gravitational collapse. Several results have been obtained recently in this direction, analyzing the gravitational collapse of dust, perfect fluid as well as collapsing shells of radiation which provide much information on the final fate of collapse and show that shell-focusing naked singularities form at the center of spherically symmetric collapsing configurations[1].

We now discuss below some of these results briefly and then consider the implications. For the sake of simplicity, we confine to the case of self-similar gravitational collapse. To begin with, consider a shell of radiation collapsing at the center in an otherwise empty space-time[10]. The gravitational potentials in (u, r, θ, ϕ) coordinates are described by the Vaidya metric given by,

$$g_{uu} = -\left(1 - \frac{2m(u)}{r}\right), \quad g_{ur} = 1, \quad g_{\theta\theta} = r^2, \quad g_{\phi\phi} = r^2 \sin^2\theta \tag{1}$$

The stress-energy tensor describes the imploding radiation and is given by

$$T_{ij} = \sigma k_i k_j \tag{2}$$

where k_i is a null vector. The collapse is characterized by the mass function $m(u)$ and the requirement of self-similarity implies that it must have a linear form,

$$m(u) = \lambda(u) \tag{3}$$

where $\lambda \geq 0$. A naked singularity forms in this case at the center $u = 0$, $r = 0$ provided $\lambda \geq 1/8$ and as shown in Ref. 10, not just isolated trajectories but families of non-spacelike trajectories with a non-zero measure are emitted from this naked singularity. Further, a powerful curvature growth is observed along all these trajectories in the limit of approach to the naked singularity where we have,

$$R_{ij} V^i V^j \propto \frac{1}{k^2} \tag{4}$$

k being the affine parameter along the geodesic. It follows that this is an essential singularity in terms of the classification of singularities.

The analysis such as above provides certain insights into what is possible in gravitational collapse in general relativity. For example, it follows that the cosmic censorship conjecture cannot be valid in the form that all naked singularities must be in some sense gravitationally weak[9]; and that the curvatures could diverge very powerfully along the non-spacelike trajectories in the limit of approach to the naked singularity.

Similar analysis can be carried out for the case of collapse of a perfect fluid as well[11], in which cases similar conclusions on the structure and formation of the naked singularity are obtained, namely, the formation of a naked singularity, emergence of a non-zero measure of non-spacelike trajectories from the same provided the positivity of the energy density holds, and a powerful curvature growth in the above sense along the non-spacelike trajectories in the limit of approach to the naked singularity in the past. In the case of a perfect fluid, the condition for the formation of the naked singularity can be written in the form of a fourth order algebraic equation which must have real positive roots.

It is possible to argue, however, that in order to generate a physical formulation for the cosmic censorship the forms of matter such as dust, perfect fluid or collapsing radiation are not really 'fundamental'. Such forms of matter must cease to be good approximations as the collapse progresses to an advanced stage. Alternatively, one could suggest[12] that these may be regarded only as approximations to more basic entities such as a massive scalar field and a massless scalar field (in the eikonal approximation). Thus, one would like to know whether naked singularities occur for a scalar field coupled to gravity, or for similar matter fields other than forms of matter such as dust, perfect fluid or collapsing radiation.

To examine this question in some detail, we consider a collapse scenario involving matter only subject to the restriction of positivity of energy but with a general equation of state[11]. The non-zero metric components for a spherically symmetric space-time are (t, r, θ, $\phi = 0, 1, 2, 3$),

$$g_{00} = -e^{2\nu},\quad g_{11} = e^{2\psi} \equiv V + X^2 e^{2\nu},\quad g_{33} = g_{22}\sin^2\theta = r^2 S^2 \sin^2\theta, \qquad (5)$$

The function V is defined as above and due to self-similarity ν,ψ,V and S are functions of the similarity parameter $X = t/r$ only. The remaining freedom in the choice of coordinates r and t can be used to set the only off-diagonal term T_{01} of the energy momentum tensor T_{ij} to zero (using comoving coordinates). We assume the matter to satisfy the weak energy condition, i.e.

$$T_{ij} V^i V^j \geq 0, \qquad (6)$$

for all non-spacelike vectors V^i. The relevant field equations for a spherically symmetric self-similar collapse of the fluid under consideration can be written as[12],

$$G_0^0 = \frac{-1}{S^2} + \frac{2e^{-2\psi}}{S}\left(X^2\ddot{S} - X^2\dot{S}\dot{\psi} + XS\dot{\psi} + \frac{(S - X\dot{S})^2}{2S}\right) - \frac{2e^{-2\nu}}{S}\left(\dot{S}\dot{\psi} + \frac{\dot{S}^2}{2S}\right) \tag{7}$$

$$= 8\pi r^2 T_0^0,$$

$$G_1^1 = \frac{-1}{S^2} + \frac{2e^{-2\nu}}{S}\left(\ddot{S} - \dot{S}\dot{\nu} + \frac{\dot{S}^2}{2S}\right) + \frac{2e^{-2\psi}}{S}\left[-SX\dot{\nu} + X^2\dot{S}\dot{\nu} + \frac{(S - X\dot{S})^2}{2S}\right] \tag{8}$$

$$G_2^2 = 8\pi T_2^2, \quad G_1^0 = \ddot{S} - \dot{S}\dot{\nu} - \dot{S}\dot{\psi} + \frac{S\dot{\psi}}{X} = T_1^0 = 0. \tag{9}$$

Using the last equation above, the first two equations can be combined to get

$$\dot{V}(X) = X e^{2\nu}[H - 2], \tag{10}$$

where $(\cdot)$ is the derivative with respect to the similarity parameter $X = t/r$ and $H = H(X)$ is defined by

$$H = r^2 e^{2\psi}(T_1^1 - T_0^0) \tag{11}$$

For matter satisfying weak energy condition, it follows that $H(X) \geq 0$ for all X. Using the field equations above and methods similar to those used in a perfect fluid collapse[11], one can see that the singularity at $t = 0$, $r = 0$ is naked when the equation $V(X) = 0$ has a real simple positive root, i.e.

$$V(X_0) = 0, \tag{12}$$

for some $X = X_0$.

It is also seen that a non-zero measure of future directed non-spacelike trajectories will escape from the singularity provided

$$0 < H_0 = H(X_0) < \infty. \tag{13}$$

Using the equations of non-spacelike geodesics in a self-similar space-time, it is seen that these escaping trajectories near the naked singularity are given by

$$r = D(X - X_0)^{\frac{2}{H_0 - 2}}. \tag{14}$$

Here D is a constant labeling different integral curves, which are the solutions of the geodesic equations, coming out of the naked singularity. It is seen that $H_0 > 0$ will hold if the weak energy condition above is satisfied and when the energy density as measured by any timelike observer is positive in the collapsing region near the singularity. In this case, when $H_0 < \infty$, families of future directed non-spacelike geodesics will come out, terminating at the naked singularity in the past. On the other hand, for $H_0 = \infty$, a single non-spacelike trajectory will come out of the naked singularity. This characterizes the formation of naked singularity in self-similar gravitational collapse. Such a singularity will be at least locally naked and considerations such as those in Ref. 11 can be used to show that it could be globally naked as well provided $V(X_0) = 0$ has more than one real simple positive roots.

We note that the existence of several classes of self-similar solutions to Einstein equations in the cases such as dust, radiation collapse etc. indicated above, where the gravitational collapse from a regular initial data results into a naked singularity, points out that such a condition will be realized for a wide variety of self-similar collapse scenarios. For example, for the case of radiation collapse with a linear mass function discussed above, the above initial condition corresponds to a restriction on the parameter λ (which is the rate of collapse) given by $0 < \lambda \leq 1/8$. In general, one could treat this condition for the existence of a naked singularity as an initial value problem for the first order differential equation given above governing $V(X)$.

One could examine the curvature strength of the naked singularity here as well, which provides an important test of its physical significance. As in the case of radiation and perfect fluid, this turns out to be a strong curvature naked singularity as specified in terms of the classification of singularities[3]. Consider the radial null geodesics coming out which are given by,

$$\frac{dt}{dr} = e^{\psi - \nu}. \tag{15}$$

Then,

$$\lim_{k \to 0} k^2 R_{ij} V^i V^j = \frac{4 H_0}{(2 + H_0)^2} > 0. \tag{16}$$

It follows that this is a strong curvature naked singularity in the sense that the volume forms defined by all possible Jacobi vector fields vanish in the limit of approach to the naked singularity, in which case the space-time may not admit any continuous extension.

Are the results such as above confined to the case of self-similar gravitational collapse only? Several results[1] worked out for the non-self-similar collapse of dust and radiation indicate that this is not the case.

The considerations such as above suggest that it may be useful to develop first a physical formulation of the cosmic censorship conjecture which captures its basic spirit before a rigorous and provable formulation for the same can be arrived at. A possibility is suggested in this direction by the results on self-similar collapse, namely that if the weak energy requirements indicated above are violated in the final stages of collapse, then even though a naked singularity may exist (in the sense that a single null geodesic might escape from the singularity), no families of non-spacelike trajectories may terminate at the naked singularity in the past. Further, this will no longer be a strong curvature singularity as described above. For all practical purposes such a naked singularity may not be taken seriously and the physical spirit of cosmic censorship hypothesis would be in tact. It is not clear what mechanism may be invoked to achieve such a violation of energy condition in the advanced stages of gravitational collapse. Of course, the quantum effects may become important in the very final stages of collapse, and the quantum gravity corrections would be relevant at extremely small lengths of the order of the Planck scales. In fact, such a possibility of violation of energy conditions has been discussed recently in the context of worm hole space-times[14].

We also learn from an analysis such as above that the formation of naked singularities in gravitational collapse is very likely not an outcome of this or that form of the matter which is collapsing, but essentially depends on the initial conditions on the physical parameters which enter the Einstein equations. It would appear that the final fate of gravitational collapse remains an issue involving exciting possibilities. Even if a successful quantum theory of gravity was some how to dispense with all the space-time singularities, one would nonetheless have to take account of essentially singular, high density regions of matter and radiation predicted by the classical theory. One would like to conclude that the phenomena of gravitational collapse, and in particular the issue of its final out come requires much more investigation than it has received so far.

References

1. P.S. Joshi (1993), 'Global aspects in gravitation and cosmology', Clarendon Press, OUP, Oxford.
2. J. Oppenheimer and R. Snyder (1939), Phys. Rev. **56**, p.455.
3. F.J. Tipler, C.J.S. Clarke and G.F.R. Ellis (1980), 'General Relativity and Gravitation' Vol. 2, ed. A Held (New York : Plenum)
4. P. Yodzis, H.-J. Seifert, H. Muller zum Hagen (1974), Commun. Math. Phys. **37**, p.29.
5. A. Papapetrou and A. Hamoi (1967), Ann. Inst. H. Poincare AVI, p.343.
6. S. Hawking (1971), Gen. Relat. Grav. Singularity Symposium.

7. D. Eardley and L. Smarr (1979), Phys. Rev. **D19**, p.2239.
8. D. Christodoulou (1984), Commun. Math. Phys. **93**, p.171
9. R.P.A.C. Newman (1986), Class. Quantum Grav. **3**, p.527.
10. I.H. Dwivedi and P.S. Joshi, Class. Quantum Grav. **6**, p.1599 (1989); Class. Quantum Grav. **8**, p.1339 (1991).
11. P.S. Joshi and I.H. Dwivedi (1992), Commun. Math. Phys. **146**, p.333; Lett. Math. Phys. **27**, p.235 (1993).
12. D.M. Eardley (1987), in 'Gravitation and Astrophysics' (Cargese, 1986), Proc. of the NATO Advanced Study Institute, Cargese, France (ed. B. Carter and J. Hartle).

Gravity Probe B: Status and Flight Plans

J.C. Mester, C.W.F. Everitt, B.W. Parkinson, and J.P. Turneaure
Hansen Experimental Physics Laboratory
Stanford University, Stanford, CA 94305 USA

Abstract

The Gravity Probe B Relativity Gyroscope Experiment will test two independent predictions of the General Theory of Relativity by measuring the precession rates of gyroscopes in a 650 km high polar orbit about the earth. The goal is to measure the geodetic effect to a precision to 2 parts in 10^5 and the frame-dragging effect to a precision of 3 parts in 10^3. This paper presents the status of the program and progress toward science mission launch.

1. Introduction

Gravity Probe B, also known as the Relativity Mission, is a space-based experiment being developed at Stanford University to test two predictions of Einstein's General Theory of Relativity. The goal of the experiment is to make very accurate measurements of the geodetic and frame-dragging effects by means of measuring changes in the spin direction of gyroscopes in polar orbit about the earth. It is an experiment that requires a dedicated satellite, which is scheduled to be launched in December 1999.

The relativistic precession, $\vec{\Omega}$ of a gyroscope in a circular orbit around the earth was calculated in 1960 by Schiff [1] and is given by:

$$\vec{\Omega} = \frac{3GM}{2c^2R^3}(\vec{R} \times \vec{\upsilon}) + \frac{GI}{c^2R^3}\left[\frac{3\vec{R}}{R^2}(\vec{R} \cdot \vec{\omega}_e) - \vec{\omega}_e\right],$$

where R is the position and υ the orbital velocity of the gyroscope, and I, M, and ω_e are the moment of inertia, mass, and angular velocity of the earth, and G is the gravitational constant.

The first term describes the geodetic precession, which arises from the curvature of space-time due to the mass of the earth. General Relativity predicts that the direction of the gyroscope will change at the rate of 6.6 arcsec per year for a gyroscope in a 650 km high polar orbit. The experimental goal is to measure the geodetic effect to 2 parts in 10^5, making this the most precise non-null test of General Relativity. The second term represents the precession due to the dragging of the inertial frame by the rotation of the earth. General Relativity predicts the rate of precession of a Gravity Probe B gyroscope to be 0.042 arcsec per year (42 marcsec/yr). The experimental goal is to measure this effect to 3 parts in 10^3. This will be the first test of General Relativity to directly measure the dragging of inertial frames by the rotation of a massive body. A polar orbit is chosen so that the two precessions are orthogonal and can therefore be discerned independently. Alternative theories of gravitation have been proposed that predict different magnitudes for either or both of these effects [2,3]. Gravity Probe B will measure the geodetic and frame-dragging effects with sufficient precision to be able to distinguish between several alternative theories and General Relativity.

2. Experiment System Overview

The small sizes of the relativity precessions require that the experiment system have extreme measurement precision and that all sources of error be controlled. In

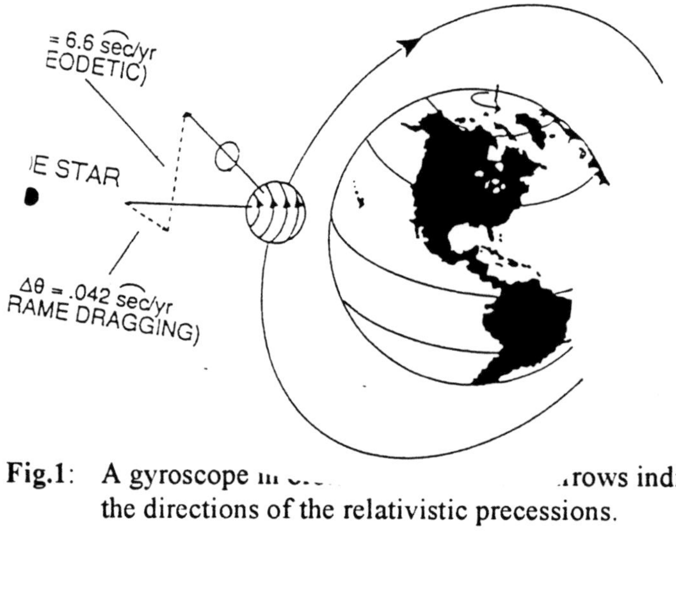

Fig.1: A gyroscope ... rows indicate the directions of the relativistic precessions.

order to achieve this requirement the experiment exploits the advantages of a near zero-g orbit in space and a near zero temperature in the experiment probe [5, 6, 7, 8]. Figure 2 shows the experiment module consisting of a helium dewar, which holds 2300 liters of superfluid helium, surrounding the experiment probe consisting of 4 gyroscopes, a quartz block, and a star tracking telescope. The dewar is designed to have an on-orbit helium lifetime of greater than 16.5 months. The helium is maintained at a temperature of 1.8 K by means of a porous plug venting system and the boil-off gas is used to power proportional thrusters used in drag-free control. The thrusters keep the spacecraft centered around a gyroscope in free fall to produce residual accelerations at this gyroscope of less than 10^{-9} g ($g = 9.8\ m/s^2$). Three other gyroscopes and a redundant drag-free proof mass are mounted within a rigid quartz block assembly. The quartz block provides precise positioning of the gyroscopes and the telescope, with cryogenic temperatures increasing mechanical stability. A series of windows provide an open line of sight out of the dewar. The star tracking telescope is used to point the spacecraft towards a guide star, providing a distant inertial reference with which to compare the gyrospin direction.

The spacecraft rolls about the line of sight to the guide star with a period of 3 minutes. This averages off-axis accelerations (which contribute to Newtonian

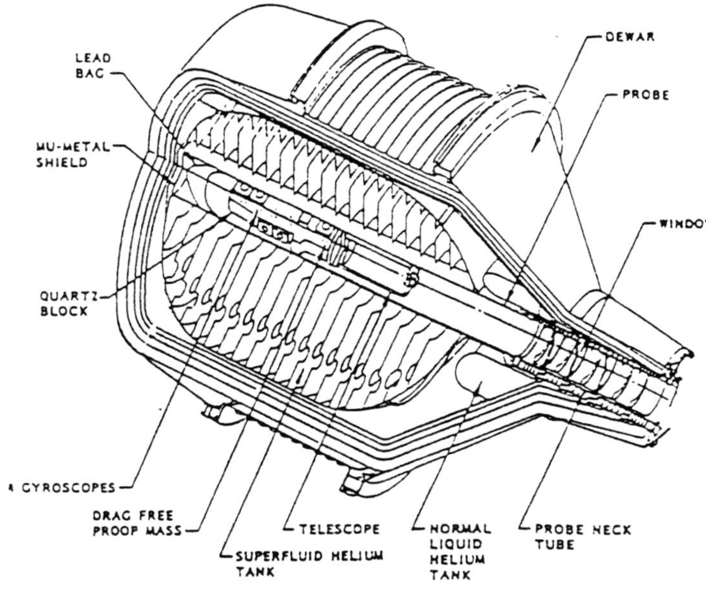

Fig.2 Gravity Probe B experiment module.

torques at the gyroscopes) to 10^{-12} g and allows the gyroscope spin direction to be measured at roll frequency, greatly reducing the effects of DC bias and drift.

3. Gyroscopes and Gyroscope Readout

The heart of the Gravity Probe B mission are the gyroscopes. The experiment accuracy dictates that the fundamental requirement for the Gravity Probe B mission is to orbit gyroscopes whose absolute Newtonian drift rates are of order or less than 0.1 marcsec/yr and whose total inertial drift can be verified to an accuracy of 0.1 marcsec/yr.. Figure 3 gives a schematic, exploded view of a gyroscope. The gyroscope is comprised of a rotor approximately 39 mm in diameter, which spins freely within the spherical cavity of a quartz housing. Newtonian torques on the gyroscope are minimized by the drag-free control system and by manufacturing rotors of high sphericity and density homogeneity.

Rotors are fabricated from fused quartz and single crystal silicon with density inhomogeneity of less than 2 parts per million and are polished to achieve peak-to-valley asphericity of less than 25 nm. The rotors are coated with a thin, uniform layer of niobium, which has a superconducting transition temperature of 9.2 K. The niobium coating enables the rotor to be electrostastically suspended within the housing and provides a means for sensing the gyroscope spin direction, discussed below. The housing for the rotor, shown in two halves split by a parting plane, has 3 orthogonal pairs of electrodes used to sense the rotor position and electrostatically suspend it. Once suspended, the rotor is spun up by sending helium gas through the spin-up channel. A final spin speed of about 130 Hz is

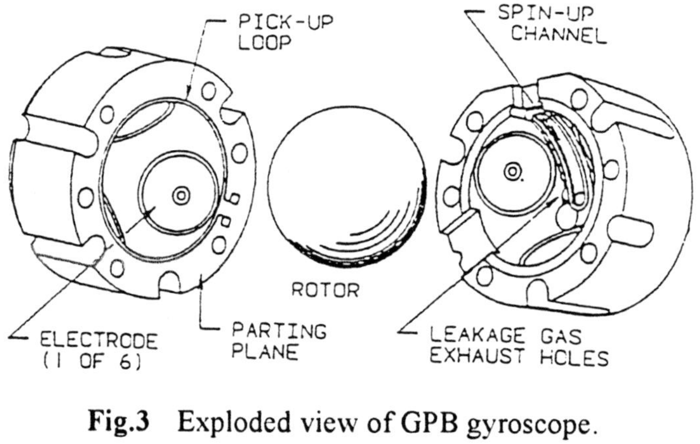

Fig.3 Exploded view of GPB gyroscope.

chosen to increase the readout signal without excessively distorting the rotor shape due to centrifugal loads. The helium gas is then pumped away to high vacuum to eliminate residual gas damping of the rotor.

As stated earlier, one of the gyroscopes is used as a drag-free sensor. No electrostatic suspension voltages are required for this gyro, which is kept centered in the housing by the positioning of the spacecraft around it. Suspension voltages are required for the other gyroscopes to overcome small accelerations ($< 10^{-7}$g) produced by the earth's gravity gradient. Suspension system forces produce the leading disturbance torques on the gyroscopes. Modeling of the gyroscope performance based on over 70,000 hours of ground testing indicates that the residual torques yield drift rates of 0.1 marcsec/yr for the suspended gyroscopes and about an order of magnitude less for the unsuspended gyroscope.

The gyroscope readout system must be capable of resolving changes in the rotor spin direction of 0.1 marcsec without producing interaction torques that could disturb that spin direction. The experiment probe's low operating temperature allows the properties of superconductivity to be exploited, both as the physical basis of the readout signal and for the signal's detection. The readout signal is based on the magnetic field produced by the London moment of a rotating superconductor [6]. As the niobium coated rotor is spun up, it develops a London magnetic moment aligned with the instantaneous spin axis. The London moment produces an equivalent magnetic field just outside the rotor of magnitude:

$$B_L = 1.14 \times 10^{-7}\,\omega_s \quad \text{Gauss,}$$

where ω_s is the spin angular velocity.

The London field is measured using a dc SQUID (superconducting quantum interference device) magnetometer, as shown in figure 4. On the parting plane between the two housing halves there is a four turn superconducting pickup loop which couples the London moment flux to the SQUID. The gyro spin axis is aligned close to the spacecraft roll axis, so the London moment produces a signal modulated at roll frequency. At our design spin speed (130 Hz), the London field is just below 1×10^{-4} Gauss. Therefore, to resolve 0.1 marcsec changes in spin direction a field sensitivity of 5×10^{-14} Gauss is required. The measured noise performance of the Gravity Probe B SQUIDs establish that 1 marsec resolution is achieved in less than 10 hours of integration time, consistent with mission requirements.

Such low field levels also dictate the need for extensive magnetic shielding. Ultralow dc fields of less than 10^{-7} Gauss, required to minimize flux trapping in

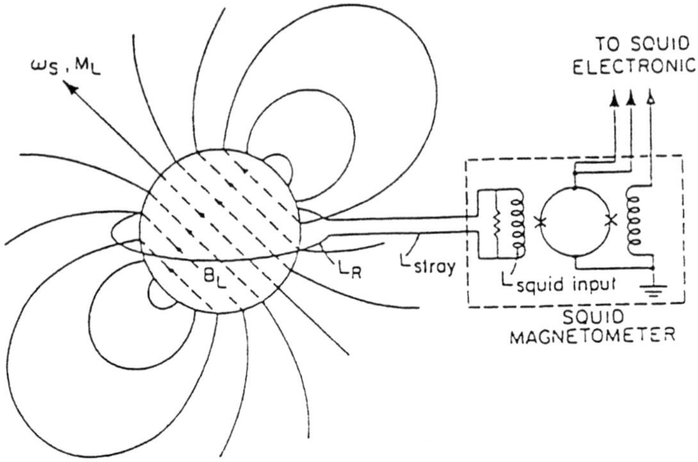

Fig.4 London moment readout system.

the rotor, are produced using the expanded superconducting lead bag shielding technique [6]. This lead shield coupled with additional internal superconducting shielding and an external cryoperm shield yield ac (roll frequency) field attenuation at the gyroscopes of greater than 1×10^{12}.

4. Telescope and Guide Star Selection

It is necessary to measure the spin direction of the gyroscopes relative to a distant reference frame, one not affected by the mass or spin of the earth. Therefore, a telescope is incorporated into the experiment module to track the position of a guide star.

The star tracking telescope is of the folded Schmidt-Cassegrainian type with a 140 mm diameter mirror and 3.81 m focal length. It is constructed out of fused quartz and has an overall physical length of 350 mm. Two focused images are formed on the edges of roop prisms by splitting the incoming star light with a beam splitter. The edges of the roof prisms are perpendicular, providing two axes of readout. Each prism divides the star image into two partial images whose intensities are determined using cryogenic silicon photo detectors and cryogenic preamplifiers. The relative intensities of the prism-split images determine the direction of the line of sight to the guide star. Using this signal, the spacecraft attitude is controlled to point in the direction of the guide star.

Cryogenic telescope tests have been performed using a ground-based light source to act as a guide star. These tests have determined that the leading noise sources, photon counting noise and telescope readout electronics noise, are well below system requirements (< 10 marcsec/$\sqrt{\text{Hz}}$) for stars brighter than magnitude 6.

An important factor in reaching measurement accuracy is the selection of the guide star to act as an inertial reference. The two main science issues concerning guide star selection are the proper motion of the guide star, or change of the guide star position with respect to very distant "fixed" stars, and the effect of declination on the magnitude of frame-dragging. For a polar orbit, Equation 1 indicates that the frame-dragging term is proportional to the cosine of the declination, thus favoring selection of a star near the equatorial plane. Uncertainties in the proper motion of the guide star contribute to experimental error and therefore the guide star proper motion needs to be known to high accuracy. Fortunately, several candidate stars brighter than the magnitude 6 threshold are also radio sources, allowing their proper motion to be determined using Very Long Baseline Interferometry (VLBI). Leading guide star candidates, HR5110, HR1099, and HR8703, now have sub or near marcsec/yr, proper motion uncertainties and VLBI observations are continuing to further reduce these uncertainties [7].

5. Ground Testing and Error Analysis

All individual experimental components are continuing to undergo intensive ground testing to provide data for error estimates. In addition, Gravity Probe B is involved in a series of rigorous integrated systems tests. We are presently proceeding with a ground test of a full scale experiment probe, with a full complement of 4 gyroscopes and a telescope, housed within an engineering dewar. This probe will serve as a backup for the science mission. The present test is investigating the interaction of all the experimental systems and is due to be completed in 1995. The science mission flight dewar is now in construction. We have scheduled an integrated test of the full scale experiment probe and the science mission flight dewar in 1997. In parallel, construction is beginning on the science mission flight experiment probe; a series of tests with the flight probe and the flight dewar (science mission payload verification) is scheduled for late 1997. The spacecraft will be completed by the end of 1998 to prepare for science mission launch.

Ground tests of the gyroscopes, telescope, and integrated systems have enabled us to determine our expected experimental error. Estimates of all individual error sources are combined to yield overall expected geodetic and frame-dragging error as a function of the mission duration. The covariant error analyses are checked against the results of Monte Carlo simulations of measurement error that include gyroscope readout system noise, telescope noise, and error in the scale factor relating the telescope and gyroscope readout directions. The most recent results predict most probable standard errors of 0.20 marcsec/yr and 0.18 marcsec/yr for the geodetic and frame-dragging measurements, respectively, for each gyroscope taken singly. For combined measurements of all four gyroscopes these errors are 0.11 marcsec/yr and 0.10 marcsec/yr, which will allow the determination of the geodetic effect to a precision of 1.7 parts in 10^5 and of the frame-dragging effect to 2.3 parts in 10^3. We anticipate mission launch from Vandenberg Air Force Base in December, 1999.

This work was supported by NASA under contract NAS8-36125. This paper summarizes the work of many members of the GP-B group at Stanford University and Lockheed Missiles and Space Systems, most notably: D. Bardas, S. Buchman, D. DeBra, J. Gwo, N. Kasdin, G. Keiser, J. Lipa, J. Lockhart, B. Muhlfelder, M. Taber, R. Van Patten, Y. Xiao, and S. Wang.

References

1. L.I. Schiff, *Proc. Nat. Acad. Sci.* **46** 871 (1960).
2. T. Damour and K. Nordtvedt, *Phys. Rev. Lett.* **70** 2217 (1993).
3. C. Will, *Theory and Experiment in Gravitational Physics* Rev. ed. Cambridge [England]; Cambridge University Press, (1993).
4. D. Bardas, *et al.*, *Proc. of SPIE* **619** 29 (1986).
5. C.W.F. Everitt, D. Davidson, and R. Van Patten, *Proc. of SPIE* **619** 89 (1986).
6. J. Lockhart, *Proc. of SPIE* **619** 148 (1986).
7. C.W.F. Everitt and S. Buchman. *XXIXth Rencontre de Moriond, Particle Astrophysics, Atomic Physics and Gravitation* Ed. by J. Tran Thanh Van, G. Fontaine, E. Hinds Editions Frontieres (1994).
8. B. Muhlfelder, *et al.*, in *Proc. of the Seventh Marcel Grossmann Meeting on General Relativity* World Scientific, Singapore, (1995).

Astro-Particle Physics in the Quasi-Steady State Cosmology

J.V. Narlikar
Inter-University Centre for Astronomy and Astrophysics,
Post Bag 4, Ganeshkhind, Pune 411 007, India

Abstract

This work highlights the conceptual and theoretical issues underlying the quasi-steady state cosmology which was proposed by the F. Hoyle, G. Burbidge and the author as an alternative to the standard big band cosmology. In particular, it is argued that this cosmology offers the high energy particle physicists several challenging problems on the cosmology-particle physics frontier.

To begin with it is shown with the help of a toy model how the problems of spacetime singularity and violation of the energy momentum conservation law that are present in the standard cosmology can be avoided by introducing a scalar field minimally coupled to gravity and having its sources in events where matter is created.

It is then shown that matter creation preferentially occurs near collapsed massive objects and the scalar field created at such mini-creation events has a feedback on spacetime geometry causing the universe to have a steady expansion as in the de Sitter model but with periodic phases of expansion and contraction superposed on it. The parameters of the toy model can be empirically fixed in relation to the cosmological observations thus providing tests of the theory.

Next it is argued that the toy model arises from a deeper theory which is Machian in origin with the inertia of a particle determined by the rest of the particles in the universe in a long range conformally invariant scalar interaction. The characteristic mass of a particle created is then the Planck mass. The Planck particle decays quickly to baryons. It is shown that the inertial effects produced by the Planck particles during their brief existence generate the scalar field of the toy model while the inertial effects of the stable baryonic particles give the more familiar Einstein equations of relativity.

The baryons into which the Planck particle decays from an SU3 octet which, in the high density - high energy environment of mini-creation event finally forms the nuclei of hydrogen, helium and other elements of low atomic masses. These predicted abundances match those actually found.

Finally it is shown that extending the theory to the most general conformally invariant form automatically leads to the cosmological constant whose sign and magnitude are of the right cosmological order.

1. Introduction

We begin with the tentative definition that cosmology refers to a study of those aspects of the universe for which spatial isotropy and homogeneity can be used, with the spacetime metric taking the form

$$ds^2 = dt^2 - S^2(t)\left[\frac{dr^2}{1-kr^2} + r^2(d\theta^2 + \sin^2\theta\, d\phi^2)\right] \tag{1}$$

in terms of coordinates t, r, θ, ϕ with $r = 0$ at the observer. The topological constant k in this so-called Robertson - Walker form can be shown to be 0 or ± 1. The "particles" to which (1) applies are thought of as galaxies or clusters of galaxies, each "particle" having spatial coordinates r, θ, ϕ independent of the universal time t. They form what is often referred to as the Hubble flow.

Big-Bang cosmology in all its forms is obtained from the equations of general relativity,

$$R_{ik} - \frac{1}{2} g_{ik} R + \lambda g_{ik} = -8\pi G T_{ik}, \tag{2}$$

which follow from the variation of an action formula

$$\mathcal{A} = \frac{1}{16\pi G}\int_V (R + 2\lambda)\sqrt{-g}\,d^4x + \int_V \mathcal{L}_{phys}(X)\sqrt{-g}\,d^4x \tag{3}$$

with respect to a general Riemannian metric

$$ds^2 = g_{ik}\ dx^i\ dx^k \tag{4}$$

within a general spacetime volume V. The physical Lagrangian $\mathcal{L}_{phys}(X)$ generates the energy-momentum tensor T_{ik} in this variation of g_{ik}. In the standard big-bang cosmology the physical lagrangian includes only particles and the electromagnetic field, whereas in inflationary forms of big-bang cosmology a scalar field is also considered to be added to $\mathcal{L}_{phys}$. This is done in various ways, being severally advocated by different authors (see Narlikar and Padmanabhan 1991 for a review).

The initial conditions assumed in the standard model are:

(i) The universe was sufficiently homogeneous and isotropic at the outset for the metric (1) to be used immediately over a range of the r-coordinate of relevance to presentday observation,

(ii) $k = 0$,

(iii) $\lambda = 0$,

(iv) The initial balance of radiation and baryonic matter was such that the light elements D, ^3He, ^4He, ^7Li were synthesised in the early universe in the following relative abundances to hydrogen

$$\frac{D}{H} \simeq \frac{^3He}{H} \simeq 2 \times 10^{-5}, \quad \frac{^4He}{H} \simeq 0.235, \quad \frac{^7Li}{H} \simeq 10^{-10} .$$

From detailed calculations these abundances can be shown to require

$$\rho_{\text{baryon}} \simeq 10^{-32} \; T^3 \; \text{gcm}^{-3}, \tag{5}$$

the radiation temperature being in degrees kelvin. (Gamow 1946, Alpher, et al 1950, 1953, Hoyle and Tayler 1964).

There is a fundamental problem here. The instant $t = 0$ is the so-called spacetime singularity at which the field equations (2) break down. This is identified with the big bang epoch. All matter that we see in the universe (as well as radiation) is supposed to be given as an initial condition at $t = \varepsilon > 0$. The initial instant ε can be taken arbitrarily close $t = 0$ but not identified with it. Thus the action principle (3) itself gets restricted in validity since the singular epoch must be excluded from it too.

Conceptually this is an exceptional step to take. In theoretical physics the basic laws or principles like the action principle are considered superior to the specific solutions based on them. Yet here we seek to restrict the validity of (2) and (3) because the solution so warrants it! There is thus a clear indication here of an inconsistency of the overall framework.

The other problems of the standard big bang model often referred to as the horizon and flatness problems also relate to the above initial conditions assumed at $t = \varepsilon > 0$. While the need for such far reaching assumptions as (i) to (iv) has always prompted a measure of unease they were widely accepted for a decade and a half, and are indeed still fully accepted by the more orthodox supporters of the standard model. Others, however, welcomed the inflationary idea of including a scalar field in the physical lagrangian that initially dominated both matter and other fields and which varied adiabatically in such a way as to give

$$\frac{\dot{S}^2}{S^2} = C, \quad S(t) = S(0)\exp\sqrt{C}\,t\,, \tag{6}$$

with C a constant. The solution (6) is considered to apply from $t = \varepsilon > 0$, where $\varepsilon \sim 10^{-36}\,s$ to a value of t large compared to $1/\sqrt{C}$. It greatly reduces the range of the r-coordinate over which (i) is needed and it effectively removes the k-term from (1). It also removes any initial contributions from matter and radiation, but these are considered to be reasserted through a physical transition of the scalar field, which jumps the solution (6) to

$$\dot{S}^2 = \frac{A}{S}\,, \quad S \simeq \left(\frac{9}{4}A\right)^{\frac{1}{3}} t^{\frac{2}{3}}\,, \tag{7}$$

which is the so-called closure model with matter just having sufficient expansion to reach a state of infinite dispersal, a condition that is considered most favourable for the eventual formation of stars and galaxies.

A major problem associated with inflation is how to effectively eliminate the cosmological constant. The value of this constant which gave the exponential solution (6) above must reduce to zero or, if the cosmological observations so demand, become as small as 10^{-108} of its initial value. Any theoretical trick invoked to achieve this has a contrived appearance (Weinberg 1989).

Last year F. Hoyle, G. Burbidge and I (1993, 1994 a,b,c) proposed an alternative scenario for cosmology called the quasi-steady state cosmology (QSSC) that gets round these conceptual difficulties as well as provides an adequate explanation of all the crucial cosmological observations. A review of the latter will be found in our previous papers (Hoyle et al op. cit.). The ideas discussed here are brief description of the theoretical aspects of our recent work.

It is commonly assumed by particle physicists that the very high energy regime that brings about a grand unification is obtainable only in the very early moments of the big bang cosmology. We show here that similar regimes exist near a typical mini-creation event of the QSSC. Moreover, unlike the 'once only never again' phase of very high energy of the big bang cosmology, here we have such phases occurring again and again, thus lending physical testability to the predictions made therein.

2. Creation of Matter : A Toy Model

The action principle (3) has a second term which is supposed to include physical contributions other than gravity. A close parallel exists between the scalar field used for inflation and the scalar field used earlier by Hoyle and Narlikar (1963) for obtaining the steady state model from Einstein's field equations. To begin with we will use the 1963 formalism as a "toy model" for describing creation of matter without violating the law of conservation of energy-momentum and without encountering spacetime singularity.

Thus the classical Hilbert action leading to the Einstein equations is modified by the inclusion of a scalar field C whose derivatives with respect to the spacetime coordinates x^i are denoted by C_i. The action is given by

$$\mathcal{A} = -\sum_a \int_{\Gamma_a} m_a \, ds_a + \int_V \frac{1}{16\pi G} R \sqrt{-g} \, d^4x - \frac{1}{2} f \int_V C_i C^i \sqrt{-g} \, d^4 x$$

$$+ \sum_a \int_{\Gamma_a} C_i \, da^i \tag{8}$$

where C is a scalar field and $C_i = \partial C / \partial x^i$. f is a coupling constant. The last term of (8) is manifestly path-independent and so, at first sight it appears to contribute no new physics. The first impression, however, turns out to be false if we admit the existence of broken worldlines. For, if particles $a, b, \ldots$ are created at world points $A_0, B_0, \ldots$ respectively, then the last term of (8) contributes a non-trivial sum

$$-\{C(A_0) + C(B_0) + \ldots\}$$

to $\mathcal{A}$.

Thus, if the worldline of particle a begins at point A_0, then the variation of $\mathcal{A}$ with respect to that worldline gives

$$m_a \frac{da^i}{ds_a} = g^{ik} C_k \tag{9}$$

at A_0. In other words, the C-field balances the energy-momentum of the created particle.

The field equations likewise get modified to

$$R_{ik} - \frac{1}{2} g_{ik} R = -8\pi G \left[\underset{m}{T_{ik}} + \underset{c}{T_{ik}} \right] \tag{10}$$

where

$$\underset{c}{T_{ik}} = -f \left\{ C_i C_k - \frac{1}{2} g_{ik} C^l C_l \right\}. \tag{11}$$

Thus the energy conservation law is

$$\underset{m}{T^{ik}}{}_{;k} = -\underset{c}{T^{ik}}{}_{;k} = fC^i C^k{}_{;k}. \tag{12}$$

That is, matter creation via a nonzero left hand side of (12) is possible while conserving the overall energy and momentum. The C-field tensor has negative stresses which lead to the expansion of spacetime, as in the case of inflation.

From (9) we therefore get a necessary condition for creation as

$$C_i C^i = m_a^2; \tag{13}$$

This is the 'creation threshold' which must be crossed for particle creation. How this can happen near a massive object, can be seen from the following simple example.

The Schwarzschild solution for a massive object M of radius $R > 2GM/c^2$ is

$$ds^2 = dt^2\left(1 - \frac{2GM}{r}\right) - \frac{dr^2}{1 - \dfrac{2GM}{r}} - r^2\,(d\theta^2 + \sin^2\theta d\,\phi^2), \tag{14}$$

for $r \geq R$. Now if the C-field does not seriously change the geometry, we would have at $r >> R$,

$$\dot{C} \approx m, \qquad C' \equiv \frac{\partial C}{\partial r} \cong 0. \tag{15}$$

If we continue this solution closer to $r \approx R$, we find that

$$C^i C_i \equiv \left(1 - \frac{2GM}{r}\right)^{-1} m^2. \tag{16}$$

In other words $C_i C^i$ increases towards the object and can become arbitrarily large if $r \approx 2GM$. So it is possible for the creation threshold to be reached *near* a massive collapsed object even if $C_i C^i$ is *below* the threshold far away from the object. In this way massive collapsed objects can provide new sites for matter creation. Further, because of the negative stresses the created matter is expelled outwards from the site while the C-field quanta escape with the speed of light. Thus, instead of a single big bang event of creation, we have mini-creation events near collapsed massive objects.

Since the C-field is a global cosmological field, we expect the creation phenomenon to be globally cophased and to have cosmological consequences. Thus, there will be phases when the creation activity is large, leading to the generation of the C-field strength in large quantities. However, the C-field growth because of its large negative stresses leads to a rapid expansion of the universe and a consequent drop in its background strength. When that happens creation is reduced and takes place only near the most collapsed massive objects thus leading to a drop in the intensity of the C-field. The reduction in C-field slows down the expansion, even leading to local contraction and so to a build-up of the C-field strength. And so on!

I emphasise a point that might be missed by particle physicists accustomed to working in a flat spacetime. When attempt is made to quantize a negative energy field, the negativity of energy leads to a run-away cascading, thus making quantization impossible. In curved spacetime, with a dynamical feedback of the field energy momentum tensor on spacetime geometry via Einstein's equations the negative energy and stresses cause expansion of space, thus providing a control on the run-away situation but also provides a feed back on the quantization process. The detailed working of the latter is still to be carried out whereas the former part is reasonably worked out.

We can describe this up and down type of activity of the mini-creation events and its impact on cosmology as an oscillatory solution superposed on a steadily expanding de Sitter type solution of the field equations as follows. For the Robertson-Walker line element the equations (10)–(12) give

$$3\,\frac{\dot{S}^2 + kc^2}{S^2} = 8\pi G\left(\rho - \frac{1}{2} f\,\dot{C}^2\right), \tag{17}$$

$$2\,\frac{\ddot{S}}{S} + \frac{\dot{S}^2 + kc^2}{S^2} = 4\pi G\, f\,\dot{C}^2, \tag{18}$$

where $S(t)$ is the scale factor and k the curvature parameter ($= 0, \pm 1$). The cosmic time is given by t. These equation have a deSitter type solution given by

$$S \propto \exp(t/P),\ k = 0,\quad \dot{C} = \text{constant},\quad \rho = \text{constant} \tag{19}$$

The oscillatory solution is given by

$$k = +1, \qquad \dot{C} \propto 1/S^3, \qquad \rho \propto 1/S^3. \tag{20}$$

Thus (17) becomes, in the latter case

$$\dot{S}^2 = -c^2 + \frac{A}{S} - \frac{B}{S^4}, \quad A, B = \text{constant}. \tag{21}$$

Here the oscillatory cycle will typically have a period $Q << P$.

Although the exact solution of (21) will be difficult to obtain, we can use the following approximate solution of (19) and (20) to describe the short-term and long-term cosmological behaviour:

$$S(t) = \exp\left(\frac{t}{P}\right)\left\{1 + \alpha \cos\frac{2\pi t}{Q}\right\}. \tag{22}$$

Note that the universe has a long term secular expanding trend, but because $|\alpha| < 1$, it also executes non-singular oscillations around it. For this reason this model has been called "quasi-steady state cosmology". We can determine α and our present epoch $t = t_0$ by the observations of the present state of the universe. Thus an acceptable set of parameters is

$$\alpha = 0.75, \quad t_0 = 0.85Q, \quad Q = 4 \times 10^{10} yr., \quad P = 20Q. \tag{23}$$

Although the set is not unique and there will be a *range* of acceptable values, we have worked with this set to illustrate the performance of model (Hoyle et al 1994 a,b) vis-a-vis observations of the universe.

What is the nature of the created particle? A deeper theory which is outlined in §4 tells us that the particle in question has Planck mass. We will assume this result to begin with and explore its consequences for astroparticle physics.

3. Problems of High Energy Physics

The Planck particle is unstable and within a timescale of $\sim 10^{-43}$ s, it decays into a large number of secondaries. The process involves a release of high energy since it begins with energy source of $\sim 10^{19}$ GeV which gets distributed over particles and radiation, the ultimate decay products being baryons, leptons and photons etc. We may see here an analogy with the descending energy ladder in the big bang cosmology, from $\sim 10^{19}$ GeV, through the GUT energy $\sim 10^{16}$ GeV, down to the electroweak unification energy of $\sim 10^{2}$ GeV, to ~ 1 GeV for baryons. Instead of a single big bang, however, we now have numerous mini-creation events involving 'Planck fireballs' centered on all decaying Planck particles.

This transition from 10^{19} GeV to 10^{2} GeV has the same range of interesting physics that particle physicists like to study in the context of the big bang cosmology. The advantage with the QSSC is that the Planck fireballs are physical objects that can be studied just like any other repetative physical phenomena. (In the 'early universe' of big bang cosmology the events are non-repetative). Moreover, many mini-creation events occur at modest redshifts $\lesssim 5$ and so are, in principle directly accessible to extragalactic astronomy, which is not the case for the early universe of big bang cosmology. As an example Narlikar and DasGupta (1993) have shown that the mini-creation events can be detected by gravity-wave detectors being planned now.

One interesting issue that is handled differently by the QSSC is the observed lack of balance between matter and antimatter. In the big bang cosmology the symmetry between matter and antimatter is normally sought to be broken during the GUTs era. Somewhat contrived scenarios are needed to understand the observed photon to baryon ratio. In the QSSC the problem is posed differently. Since the universe 'renews itself' over a few oscillations, we have to understand *why*, given a matter dominated phase now, it will persist even with the renewed phase. Since the *C*-field is a globally interacting field the imbalance in the current phase is expected to be propagated into the next. How exactly the propagation of broken symmetry takes place is still to be worked out.

Finally, let us see what happens at the limit of the decay process, when from the Planck particle we end up with a group of baryons and radiation. At temperature >> 1 Ge V we expect an equipartition of all eight particles of the baryon octet. Eventually, however, all except the more long-lived neutron and proton decay to proton and end up as hydrogen nuclei. The neutron and proton combine to form the helium nuclei. A simple counting thus tells us that with two out of eight particles forming helium we expect the helium abundance to be ~ 0.25 by mass.

A more detailed calculation has been given by Hoyle et al (1993) and it leads to values of not only the helium abundance but also of D, Li, Be, B including their isotopes that agree with observations. The important difference is that instead of the assumption (iv) and equations (5) of §1, we have here a different set of values for ρ and T with the result that the deuterium abundance does not constrain the cosmological baryonic matter density. In other words, dark matter can be baryonic. This issue therefore has implications not only for astrophysics and cosmology (Hoyle et al a,b) but also for particle physics.

4. Scale-Invariant Gravity

An important property of physical theories is scale invariance or conformal invariance. Maxwell's equations and the Dirac equation for massless particle are conformally invariant but general relativity is not. If, however, the inertial mass transforms inversely as the length scale in conformal transformation then the Dirac equation for a massive fermion as well as classical and quantum electrodynamics will become conformally invariant. Can general relativity be suitably reformulated to become conformally invariant? We indicate the steps towards this goal since they naturally lead to comprehensive theory of matter creation that encompasses the toy model described above.

It is necessary to begin by finding an action $\mathcal{A}$ that is unaffected in its value by a scale transformation. The second term on the right-handed side of (3) can be made to satisfy this requirement. For a set of particles $a, b, \ldots$ of masses $m_a, m_b, \ldots$ the form of $\mathcal{L}_{\text{phys}}$ usually considered in gravitational theory is

$$\sum_{a, b, \ldots} \int \frac{\delta_4 (X, A)}{\sqrt{-g(A)}} m_a (A) \, da, \tag{24}$$

where the possibility of the particle masses varying with the spacetime position requires the mass $m_a(A)$ of particle a to vary with the point A on its path, and similarly for the other particles. Hence the second term on the right-hand side of (3) is

$$-\sum_a \int m_a (A) \, da. \tag{25}$$

with $da^* = \Omega \, da$ and $m_a^* = \Omega^{-1} m_a$ it is clear that (25) is invariant with respect to a conformal (scale) transformation.

The possibility of particle masses varying with spacetime coordinates arises most naturally in a Machian approach. Here the property of inertia is not entirely intrinsic to a particle but is also related to its presence in a non-empty universe. A quantitative description of this idea that we will follow here is based on an early work by two of us (Hoyle and Narlikar 1964). In this inertia is expressed as a scalar conformally invariant long range interaction between particles.
To begin with choose a scalar mass field $M(X)$ to satisfy

$$\Box_X M(X) + \frac{1}{6} RM(X) = \sum_a \int \frac{\delta_4 (X, A)}{\sqrt{-g(A)}} \, da \tag{26}$$

Equation (26) has both advanced and retarded solutions. We particularize an advanced solution $M^{\text{adv}}(X)$ and a retarded solution $M^{\text{ret}}(X)$ in the following way. $M^{\text{ret}}(X)$ is to be the so-called fundamental solution in the flat spacetime limit (Courant and Hilbert, 1962). This removes for $M^{\text{ret}}(X)$ the ambiguity that would obviously arise from the homogeneous wave equation. The corresponding ambiguity for $M^{\text{adv}}(X)$ is removed by the physical requirement that fields without sources are to be zero. Since

$$\Box\,[M^{\text{adv}} - M^{\text{ret}}] + \frac{1}{6} R\,[M^{\text{adv}} - M^{\text{ret}}] = 0, \tag{27}$$

the immediate consequences of this boundary condition is that $M^{\text{adv}} - M^{\text{ret}}$, being without sources, must be zero, so that

$$M^{\text{adv}}(X) = M^{\text{ret}}(X) = M(X) \text{ say.} \tag{28}$$

The gravitational equations are now obtained by putting

$$m_a(A) = M(A), \quad m_b(B) = M(B), \ldots \tag{29}$$

It can also be shown that in a conformal transformation the mass field $M(X)$ transforms as

$$M^*(X) = \Omega^{-1}(X) M(X), \tag{30}$$

a result that follows from the form of the wave equation (10) (c.f. Hoyle and Narlikar, 1974, 111). The outcome *(loc. cit., 112 et seq)* is

$$K\left(R_{ik} - \frac{1}{2} g_{ik} R\right) = -T_{ik} + M_i M_k - \frac{1}{2} g_{ik} g^{pq} M_p M_q + g_{ik} \Box K - K_{;ik}, \tag{31}$$

where

$$K = \frac{1}{6} M^2. \tag{32}$$

These gravitational equations are scale invariant. It may seem curious that from a similar beginning, (24) for the action rather than (3), the outcome is more complicated, but this seems to be a characteristic of the physical laws. As the laws are improved they become simpler and more elegant in their initial statement but more complicated in their consequences.

Now make the scale change

$$\Omega(X) = M(X)/\tilde{m}_0, \tag{33}$$

where $\tilde{m}_0$ is a constant with the dimensionality of $M(X)$. After the scale change the particle masses simply become $\tilde{m}_0$ everywhere and in terms of transformed masses the derivative terms drop out of the gravitational equations. And defining the gravitational constant G by

$$G = \frac{3}{4\pi \tilde{m}_0^2}, \tag{34}$$

the equations (31) take the form of general relativity

$$R_{ik} - \frac{1}{2} g_{ik} R = - 8\pi G T_{ik} \tag{35}$$

It now becomes clear why the equations of general relativity are not scale invariant. These are the special form to which the scale invariant equations (31) reduce with respect to a particular scale, namely that in which particle masses are everywhere the same.

It is also clear that the transition from (31) to (35) is justified provided $\Omega(X) \neq 0$ or $\Omega(X) \neq \infty$. For example, if $M(X) = 0$ on a spacelike hypersurface the above conformal transformation breaks down. It is because of the existence of such time sections that the use of (35) leads to the (unphysical) conclusion of a spacetime singularity. It has been shown [Hoyle and Narlikar 1974, Kembhavi 1979] that the various spacetime singularities like that in the big bang or in a black hole collapse arise because of the illegitimate use of (35) in place of (31).

It is easily seen from the wave equation (26) that $M(X)$ has dimensionality $(\text{length})^{-1}$, and so has $\tilde{m}_0$. Units are frequently used in particle physics for which both the speed of light c and Planck's constant $\hbar$ are unity and in these units mass has dimensionality $(\text{length})^{-1}$. If we suppose these units apply to the above discussion then from (34)

$$\tilde{m}_0 = (3/4\pi G)^{1/2}, \tag{36}$$

which with $c = \hbar = 1$ is the mass of the Planck particle. This suggests that in a gravitational theory without other physical interactions the particles must be of mass (36), which in ordinary practical units is about 10^{-5} gram, the empirically determined value of G being used. This conclusion can be supported at greater length [See Hoyle, et al 1944c]. We will, however, next consider what happens when the Planck mass decays into the much more stable baryons.

5. The Creation of Matter : A Machian Description

A typical Planck particle a exists from A_0 to $A_0 + \delta A_0$, in the neighbourhood of which it decays into n stable baryonic secondaries, $n \simeq 6.10^{18}$, denoted by $a_1, a_2, \ldots a_n$. Each such secondary contributes a mass field $m^{(a_r)}(X)$, say, which is the fundamental solution of the wave equation

$$\Box\, m^{(a_r)} + \frac{1}{6} R m^{(a_r)} = \frac{1}{n} \int_{\sim A0 + \delta A0} \frac{\delta_4 (X, A)}{\sqrt{-g(A)}}\, d a, \tag{37}$$

while the brief existence of a contributes $c^{(a)}(X)$, say, which satisfies

$$\Box c^{(a)} + \frac{1}{6} R c^{(a)} = \int_{A0}^{A0 + \delta A0} \frac{\delta_4 (X, A)}{\sqrt{-g(A)}}\, da. \tag{38}$$

Summing $c^{(a)}$ with respect to $a, b, \ldots$ gives

$$c(X) = \sum_a c^{(a)} (X), \tag{39}$$

the contribution to the total mass $M(X)$ from the Planck particles during their brief existence, while

$$\sum_a \sum_{r=1}^{n} m^{(a_r)} (X) = m(X) \tag{40}$$

gives the contribution of the stable particles.

Although $c(X)$ makes a contribution to the total mass function

$$M(X) = c(X) + m(X) \tag{41}$$

that is generally small compared to $M(X)$, there is the difference that, whereas $m(X)$ is an essentially smooth field, $c(X)$ contains small exceedingly rapid fluctuations and so can contribute significantly to the derivatives of $c(X)$. The contribution to $c(X)$ from Planck particles a, for example, is largely contained between two light cones, one from A_0, the other from $A_0 + \delta A_0$. Along a timelike line cutting these two cones the contribution to $c(X)$ rises from zero as the line crosses the light cone from A_0, attains some maximum value and then falls back effectively to zero as the line crosses the second light cone from $A_0 + \delta A_0$. The time derivative of $c^{(a)}(X)$ therefore involves the reciprocal of the time difference between the two light cones. This reciprocal cancels the short duration of the source term on the right-hand side of (40). The factor in question is of the order of the decay time τ of the Planck particles, $\sim 10^{-43}$ seconds. No matter how small τ may be the reduction in the source strength of $c^{(a)}(X)$ is recovered in the derivatives of $c^{(a)}(X)$, which therefore cannot be omitted from the gravitational equations.

The derivatives of $c^{(a)}(X)$, $c^{(b)}(X) \ldots$ can as well be negative as positive, so that in averaging many Planck particles, linear terms in the derivatives do disappear. It is therefore not hard to show that after such an averaging the gravitational equations become

$$R_{ik} - \frac{1}{2} g_{ik} R = \frac{6}{m^2}\Bigg[- T_{ik} + \frac{1}{6}(g_{ik}\Box m^2 - m^2_{;ik}) + (m_i m_k - \frac{1}{2} g_{ik} m_l m^l)$$

$$+ \frac{2}{3}(c_i c_k - \frac{1}{4} g_{ik} c_l c^l)\Bigg]. \quad (42)$$

Since the same wave equation is being used for $c(X)$ as for $m(X)$, the theory remains scale invariant. A scale change can therefore be introduced that reduces $M(X) = m(X) + c(X)$ to a constant, or one that reduces $m(X)$ to a constant. Only that which reduces $m(X)$ to a constant, viz

$$\Omega = \frac{m(X)}{m_0} \quad (43)$$

has the virtue of not introducing small very rapidly varying ripples into the metric tensor. Although small in amplitude such ripples produce non-negligible contributions to the derivatives of the metric tensor, causing difficulties in the evaluation of the Riemann tensor, and so are better avoided. Simplifying with (43) does not bring in this difficulty, which is why separating of the main smooth part of $m(X)$ in (41) now proves an advantage, with the gravitational equations simplifying to

$$8\pi G = \frac{6}{m_0^2}, \quad m_0 \text{ a constant,} \quad (44)$$

$$R_{ik} - \frac{1}{2} g_{ik} R = - 8\pi G\left[T_{ik} - \frac{2}{3}\left(c_i c_k - \frac{1}{4} g_{ik} c_l c^l\right)\right]. \quad (45)$$

Using the metric (1) with $k = 0$ the dynamical equations for the scale factor $S(t)$ are

$$\frac{2\ddot{S}}{S} + \frac{\dot{S}^2}{S^2} = \frac{4\pi}{3} G \bar{\dot{c}}^2, \quad (46)$$

$$\frac{3\dot{S}^2}{S^2} = 8\pi G\left(\bar{\rho} - \frac{1}{2}\bar{\dot{c}}^2\right), \quad (47)$$

with $\bar{\rho}$ the average particle mass density and $\bar{\dot{c}}^2$ being the average value of $\dot{c}^2$, the average value of terms linear in c and of $\ddot{c}$ being zero. It is easily shown from (46) and (47) that

$$\frac{\partial \bar{\rho}}{\partial t} + \frac{3\dot{S}}{S}\bar{\rho} = \frac{1}{2}\left(\frac{\partial \bar{\dot{c}}^2}{\partial t} + \frac{4\dot{S}}{S}\bar{\dot{c}}^2\right). \tag{48}$$

If at a particular time there is no creation of matter then at that time the left-hand side of (48) is zero with $\bar{\rho} \propto S^{-3}$. And with the right-hand side also zero at that time $\bar{\dot{c}}^2 \propto S^{-4}$. The sign of the $\dot{c}^2$ term in (46) is that of a negative pressure, a characteristic of the fields introduced into inflationary cosmological models. The concept of Planck particles forces the appearance of a negative pressure. In effect the positive energy of created particles is compensated by the sign of the $\bar{\dot{c}}^2$ terms, which in (46) increases $\ddot{S}/S$ and so causes the universe to expand. One can say that the universe expands because of the creation of matter. The two are connected because the divergence of the right-hand side of the gravitational equations (45) is zero.

As would be expected from this conservation property the sign of the $\bar{\dot{c}}^2$ term in (47) is that of a negative energy field. Such fields have generally been avoided in physics because in flat spacetime they would produce catastrophic instabilities - creation of matter with positive energy producing a negative energy $\bar{\dot{c}}^2$ term, producing more matter, producing a still larger $\bar{\dot{c}}^2$ term, and so on. Here the effect is to produce explosive outbursts from regions where any such instability takes hold, through the $\dot{c}^2$ term in (46) generating a sharp increase of $\ddot{S}$. The sites of the creation of matter are thus potentially explosive. The explosive expansion of space serves to control the creation process and avoids the catastrophic cascading down the negative energy levels.

The requirement is in agreement with observational astrophysics which in respect of high energy activity is all of explosive outbursts, without evidence for the ingoing motions required by the popular accretion-disk theory for which there is no direct observational support. The profusion of sites where X-ray and γ - ray activity is occurring are on the present theory sites where the creation of matter is currently taking place.

A connection with the toy model can now be given. Writing

$$C(X) = \tau\, c(X), \tag{49}$$

where τ is the decay lifetime of the Planck particle, the action contributed by Planck particles $a, b, \ldots,$

$$-\sum_a \int_{A0}^{A0+\delta A0} C(A)\, da \tag{50}$$

can be approximated as

$$-C(A_0)-C(B_0)-\ldots, \tag{51}$$

which form was used in the toy model. And the wave-equation for *C(X)*, using the same approximation, is

$$\Box C+\frac{1}{6}RC=\tau^{-2}\sum_a\frac{\delta_4(X,A_0)}{\sqrt{-g(A_0)}}, \tag{52}$$

which was also used in the toy model, except that the 1 / 6 *RC* term is included in the wave equation and previously an unknown constant *f* appeared in place of τ^2.

6. A Derivation of the Cosmological Constant

Writing $M^{(a)}(X), M^{(b)}(X), \ldots$ as the mass fields produced by the individual Planck particles *a,b*, . . ., the total mass field

$$M(X)=\sum_a M^{(a)}(X) \tag{53}$$

satisfies the wave equation (26) when $M^{(a)}$, $M^{(b)}$, . . . satisfy

$$\Box M^{(a)}+\frac{1}{6}RM^{(a)}=\int\frac{\delta_4(X,A)}{\sqrt{-g(A)}}\,da,\ldots \tag{54}$$

Scale invariance throughout requires all the mass fields to transform as

$$M^{*(a)}=M^{(a)}\Omega^{-1} \tag{55}$$

with respect to the scale change Ω, when both the left and right hand sides of every wave equation transform to its starred form multiplied by Ω^{-3}, i.e. the left hand side of (54) goes to $(\Box M^{*(a)}+\frac{1}{6}R^{*}M^{*(a)})\,\Omega^{-3}$ and the right hand side to

$$\Omega^{-3}\int\frac{\delta_4(X,A)}{\sqrt{-g^{*}(A)}}\,da^{*}. \tag{56}$$

Then the factor Ω^{-3} cancels to give the appropriate invariant equation. This cancellation is evidently unaffected if, instead of (54) for the wave equation satisfied by M^{a}, we have

$$\Box M^{(a)} + \frac{1}{6} RM^{(a)} + M^{(a)^3} = \int \frac{\delta_4 (X, A)}{\sqrt{-g(A)}} da. \qquad (57)$$

Since the cube term transforms to $M^{*(a)^3} \Omega^{-3}$ with respect to Ω changing (54) to (57) preserves scale invariance in what appears to be its widest form. Since in other respects the laws of physics always seem to take on the widest ranging properties that are consistent with the relevant forms of invariance we might think it should also be here, in which case (57) rather than (54) is the correct wave equation for $M^{(a)}$, and similarly for $M^{(b)}$, . . ., the mass fields of the other Planck particles.

But this departure from linearity in the wave equations for the individual particles prevents a similar equation being obtained for $M = \sum_a M^{(a)}$. Nevertheless, the addition of the individual equations can be considered in a homogeneous universe to lead to an approximate wave equation for M of the form

$$\Box M + \frac{1}{6} RM + \Lambda M^3 = \sum_a \int \frac{\delta_4 (X, A)}{\sqrt{-g(A)}} da, \qquad (58)$$

$$\Lambda = N^{-2} \qquad (59)$$

where N is the effective number of particles contributing the sum $\sum_a M^{(a)}$. The latter can be considered to be determined by an Olbers-like cut-off, contributed by the portion of the universe surrounding the point X in $M(X)$ to a redshift of order unity. In the observed universe this total mass $\sim 10^{22}\, M_\odot$, sufficient for $\sim 2.10^{60}$ Planck particles. The actual particles are of course nucleons of which there are $\sim 10^{79}$. But if suitably aggregated they would give $\sim 2.10^{60}$ Planck particles and with this value of N

$$\Lambda \simeq 2.5 \times 10^{-121}. \qquad (60)$$

The next step is to notice that the wave-equation (58) would be obtained in usual field theory from $\delta A = 0$ for $M \rightarrow M + \delta M$ applied to

$$\mathcal{A} = -\frac{1}{2} \int \left(M_i M^i - \frac{1}{6} RM^2 \right) \sqrt{-g}\, d^4x + \frac{1}{4} \Lambda \int M^4 \sqrt{-g}\, d^4x$$

$$- \sum_a \int \frac{\delta_4 (X, A)}{\sqrt{-g\,(A)}} M(X)\, da. \qquad (61)$$

In the scale in which M is m_0 everywhere the derivative term in (61) vanishes and since $G = 3/4\pi\, m_0^2$ the term in R is the same as in (3), as are also the line integrals, requiring the remaining term to be the same gives

$$\lambda = -3\,\Lambda\, m_0^2\,. \tag{62}$$

Thus we have obtained not only a cosmological constant but also its magnitude, something that lies beyond the scope of the usual theory. With 2.5×10^{-121} for Λ as in (60) and with m_0 the inverse of the Compton wavelength of the Planck particle, $\sim 3.10^{32}\, cm^{-1}$, (62) gives

$$\lambda \simeq -2.10^{-56}\, cm^{-2}, \tag{63}$$

agreeing closely with the magnitude that has previously been assumed for λ. In the classical big bang cosmology there is no dynamical theory to relate the magnitude of λ to the density or other physical properties of matter. For observational consistency it is assumed that λ is of order (63). A dynamical derivation is possible if one goes into the very early inflationary epochs. However, the values of λ deduced from those calculations are embarrassingly large, being $10^{108} - 10^{120}$ times the value given by (63). The problem then is, how to reduce λ from such high values to the presently acceptable range (Weinberg 1989). By contrast, the present derivation leads to the acceptable range of values with very few theoretical assumptions.

7. Conclusions

The theory developed in this paper differs from big-bang cosmology in what we believe to be an important aspect, that the gravitational equations are scale invariant. The gravitational equations including both the creation terms and the cosmological constant then reduce in the constant mass frame to

$$R_{ik} - \frac{1}{2} g_{ik} R + \lambda\, g_{ik} = -8\pi G \left[T_{ik} - \frac{2}{3}\left(c_i c_k - g_{ik} \frac{1}{4} c_l\, c^l \right)\right]. \tag{64}$$

The immediate successes of the theory are :

(i) The circumstance that G determined by (34) is necessarily positive requires gravitation to act as an attractive force. Aggregates of matter must tend to pull together. This is unlike general relativity where gravitation can as well be centrifugal, with aggregates of matter blowing always apart, as follows if G in the action (3) of general relativity is chosen to be negative.

(ii) In the cosmological case with homogeneity and isotropy the pressure contributed by the c-field term in the gravitational equations is negative, explaining the expansion of the universe.

(iii) Also in the cosmological case, the energy contribution of the c-field is negative, which ensures that when the creation condition (9) are satisfied the creation process tends to cascade with explosive consequences.

(iv) The magnitude of the constant λ is shown to be of the order needed for cosmology. Unlike big-bang cosmology this is a deduction not an assumption.

There are also a few unsolved questions relating to particle physics which are briefly stated thus :

(i) How is the theory quantized in curved spacetime, including its dynamical feedback on spacetime geometry?

(ii) How do we understand the particle-antiparticle asymmetry as an effect propagated (through a global interaction) from one generation of oscillation to the next?

(iii) Do we need non-baryonic dark matter at all?

(iv) Can we relate the parameters (23) to fundamental physical constants?

We hope that, given the successes achieved by the empirical model in explaining the cosmological observations, we can interest at least a few particle physicists in the attempts to come to grips with the solution of the above problems.

References

1. Alpher, R.A., Follin, J.W. and Herman, R.C., 1953, *Phys. Rev.*, **92**, 1347.
2. Alpher, R.A., Follin, J.W. and Herman, R.C., 1950, *Rev. Mod. Phys.*, **22**, 153.
3. Courant and Hilbert, D., 1962, *Methods of Mathematical Physics, Vol. II* (Interscience, New York), p. 727-744.
4. Gamow, G., 1946, *Phys. Rev.*, **70**, 572.
5. Hoyle, F., Burbidge, G.R. and Narlikar, J.V., 1993 *Ap. J.*, **410**, 437
6. Hoyle, F., Burbidge, G.R. and Narlikar, J.V., 1994a, M.N.R.A.S., **267**, 1007.
7. Hoyle, F., Burbidge, G.R. and Narlikar, J.V., 1994b, A & A, to be published.

8. Hoyle, F., Burbidge, G.R. and Narlikar, J.V., 1994c, *Proc. R. Soc.* **A**, to be published.
9. Hoyle, F. and Narlikar J.V., 1963, *Proc. R. Soc.* **A 273**, 1.
10. Hoyle, F. and Narlikar J.V., 1964, *Proc. R. Soc.* **A 282**, 191.
11. Hoyle, F. and Narlikar J.V., 1974, *Action at a Distance in Physics and Cosmology* (W.H. Freeman, New York).
12. Hoyle, F. and Tayler, R.J., 1964, *Nature*, **203**, 1108.
13. Kembhavi, A.K., 1979, M.N.R.A.S., **185**, 807.
14. Narlikar, J.V. and DasGupta, P., 1993, M.N.R.A.S, **264**, 489.
15. Narlikar, J.V. and Padmanabhan, T., 1991, *Ann. Rev. A & A*, **29**, 325.
16. Weinberg, S., 1989, *Rev. Mod. Phys.*, **61**, 1.

Field Singularities in General Relativity

J. Krishna Rao
Department of Mathematics
Bhavnagar University
Bhavnagar - 364 002

Abstract

The occurrence of matter singularities in gravitational collapse of stars has been discussed extensively in the literature. But a more serious nature of singularity which occurs in general relativity is due to the gravitational field itself, the part represented by the Weyl conformal tensor. Examples of field singularities due to spherically symmetric distributions of matter with (i) uniform (including constant) density (ii) shear free motion of charged fluids with uniform matter density and (iii) stiff fluid are discussed. In the first two cases the Weyl conformal curvature falls off as inverse cube of distance from the centre whereas in the third case it falls off as inverse square showing that the gravitational energy stored in the field dominates over the material energy as the singularity is approached.

1. Introduction

Detailed investigations about the ultimate fate of a gravitating system, both in the local and global context, has led to the cosmic censorship conjecture (Penrose, 1969) which abhorred the existence of naked singularities. The occurrence of strong curvature singularities in the pure radiation filled Vaidya's (1943, 1951, 1952, 1953) space-time as well as the dust filled Tolman (1934a, 1949) and Bondi (1947) class of inhomogeneous models arose doubts about the efficacy of the cosmic censorship conjecture. However, since the matter filling these models is not considered physically realistic the cosmic censorship conjecture continues to hold its sway. While the existence of naked singularities has brought out the limitations of Einstein's gravitation theory, a still more serious draw back of the theory, that is, the existence of *field singularities* did not receive the same attention. In the present paper we cite some examples where such field singularities do occur.

It is well known that the Riemann curvature tensor can be decomposed into the matter (as represented by the Ricci tensor) and free gravitational field (as represented by the conformal Weyl tensor) parts. The general Bianchi identities describe the interaction between the matter and free gravitational fields. In the present paper, assuming spherical symmetry, and material distribution of both uniform (including constant) and non-uniform density, we show that the free gravitational field produces a singularity at the centre. Further we show that in the case of a non-shearing charged fluid of uniform density filling the sphere, the combined electromagnetic and free gravitational fields produce a singularity at the centre.

In Section 2 we consider the well known static Schwarzschild interior solution and identify the constant of integration occurring in the radial component of the metric tensor as related with the conformal Weyl tensor (free gravitational field.) A more general situation where the sphere is filled with matter of uniform density is considered in Section 3 and the case of a charged sphere of uniform material density was discussed in Section 4. In Section 5 we discuss spherical models filled with a stiff fluid and concluding remarks are given in Section 6.

2. The Schwarzschild Interior Solution

The metric for a spherically symmetric static space-time is written using the curvature coordinates r, θ, ϕ and time t as

$$ds^2 = -\exp\lambda\, dr^2 - r^2\,(d\theta^2 + \sin^2\theta\, d\phi^2) + \exp\nu\, dt^2, \tag{1}$$

where λ and ν are functions of r only. Assuming that the space-time described by (1) is filled with a perfect fluid of proper pressure $p(r)$ and density $\rho(r)$, we obtain

$$8\pi p = \exp(-\lambda)\ \{(\nu'/r) + (1/r^2)\}\ - (1/r^2), \tag{2}$$

$$8\pi\rho = \exp(-\lambda)\ \{(\lambda'/r) - (1/r^2)\}\ + (1/r^2), \tag{3}$$

and the conservation equations reduce to

$$-dp/dr = (\rho + p)\,\nu'/2, \tag{4}$$

where a prime for λ and ν denotes a differentiation with respect to r.

Further, if the material density ρ is assumed to be a constant, as in the case of the Schwarzschild interior solution, we can immediately integrate (3) to obtain the expression for $\exp(-\lambda)$ in the form (Tolman, 1934 b).

$$\exp(-\lambda) = 1 - (8\pi\rho/3)\,r^2 + (A/r). \tag{5}$$

A being a constant of integration. To avoid a singularity at the origin it is generally assumed that $A = 0$. However, we shall show that A is not just a constant of integration but the term (A/r) producing a singularity at the centre of the sphere is related to the free gravitational field, the part represented by the Weyl conformal tensor. For this purpose we shall introduce the eigen value ε of the Weyl conformal tensor (Krishna Rao 1966) for the space-time (1) into the transverse components of the Einstein tensor ($G_2^2 = G_3^3$) so that the expression for pressure takes the alterative form (Krishna Rao and Annapurna, 1985)

$$8\pi p = -8\pi\varepsilon + \exp(-\lambda)\left[\{(\nu' - \lambda')/r\} - (1/r^2)\right] + (1/r^2), \tag{6}$$

where

$$8\pi\varepsilon = -\exp(-\lambda)\left[(\nu''/2) + \{(\nu' - \lambda')(\nu' r - 2)/4r\} - (1/r^2)\right] + (1/r^2), \tag{7}$$

denotes the only real eigen value of the conformal Weyl tensor in the classification of gravitational fields (Petrov, 1954, Pirani, 1957, 1962 a, b).

We now use the combination $\{(2) + (3) - (6)\}$ to obtain $\exp(-\lambda)$ so that

$$\exp(-\lambda) = 1 - 8\pi(\rho + \varepsilon) r^2/3, \tag{8}$$

and since ε is coupled to ρ we call the former, *the energy density of the free gravitational field (Krishna Rao, 1972a)*. We may mention here that in (6), for simplicity, we have taken the coupling constant as unity.

For the Schwarzschild interior case, we compare (8) with (5) to obtain the expression for ε as

$$\varepsilon = 3A/8\pi r^3. \tag{9}$$

We know that $\rho \propto r^{-3}$ and the expression for ε given by (9) also shows the typical r^{-3} fall of for the type D (Coulomb) gravitational field. In the next section we shall generalize this result to the case of a sphere filled with uniform density.

3. Fluid spheres of uniform density

For the discussion of spheres filled with a material of uniform density, viz. $\rho = \rho(t)$, it is convenient to write the metric using the Eulerian coordinate $R(r, t)$ so that $R(r, 0) = r$. Thus, the space-time metric takes the form

$$ds^2 = -\exp\lambda\, dr^2 - R^2(d\theta^2 + \sin^2\theta\, d\phi^2) + \exp\nu\, dt^2, \tag{10}$$

where λ, R and ν are functions of r and t. Here, we identify r as the comoving coordinate so that the velocity vector u^a of the fluid is written as $u^a = \{0, 0, 0, \exp(-\nu/2)\}$. The non-vanishing kinematical quantities associated with u^a are given by

$$\theta = \exp(-\nu/2)\ \{(\dot{\lambda}/2) + (2\dot{R}/R)\}, \tag{11}$$

$$3\sigma^1_1/2 = -3\sigma^2_2 = -3\sigma^3_3 = \exp(-\nu/2)\ \{(\dot{\lambda}/2) - (\dot{R}/R)\}, \tag{12}$$

where θ denotes the expansion (or contraction) and σ^b_a the shear tensor.

Thus, the expressions for the material energy density ρ, pressure p are expressed in the form of Einstein's equations as

$$8\pi T^1_1 = -8\pi p = -[1/R\exp(\lambda)]\ \{(R'^2/R) + R'\nu'\}$$
$$+[1/R\exp(\nu)]\ \{2\ddot{R} + (\dot{R}^2/R) - \dot{R}\dot{\nu}\} + (1/R^2), \tag{13}$$

$$8\pi T^2_2 = 8\pi T^3_3 = -8\pi p = 8\pi\varepsilon$$
$$-[1/R\exp(\lambda)]\ \{2R'' - (R'^2/R) - R'\lambda' + R'\nu'\}$$
$$+[1/R\exp(\nu)]\ \{2\ddot{R} - (\dot{R}^2/R) + \dot{R}\dot{\lambda} - \dot{R}\dot{\nu}\} - (1/R^2) \tag{14}$$

$$8\pi T^4_4 = 8\pi\rho = -[1/R\exp(\lambda)]\ \{2R'' + (R'^2/R) - R'\lambda'\}$$
$$+[1/R\exp(\nu)]\ \{(\dot{R}^2/R) + \dot{R}\dot{\lambda}\} + (1/R^2), \tag{15}$$

$$8\pi T^1_4 = 0 = [1/R\exp(\lambda)]\ (2\dot{R}' - R'\dot{\lambda} - \dot{R}\nu'), \tag{16}$$

where

$$8\pi\varepsilon = \exp(-\lambda)\ [(R''/R) - (R'/R)^2 - (\nu''/2) - (\nu'^2/4)$$
$$-(R'\lambda'\ /2R) + (R'\nu'/2R) + (\lambda'\nu'/4)]$$
$$+\exp(-\nu)[(\ddot{\lambda}/2) + (\dot{\lambda}^2/4) - (\ddot{R}/R) + (\dot{R}/R)^2$$
$$-(\dot{R}\dot{\lambda}/2R) + (\dot{R}\dot{\nu}/2R) - (\dot{\lambda}\dot{\nu}/4)] + (1/R^2), \tag{17}$$

is the real eigen value of the conformal Weyl tensor for the non-static space-time (10). Here and in what follows a prime and an overhead dot denote respectively a differentiation with respect to r and t.

Once again, making the combination {(13) + (15) − (12)}, we obtain the expression for $\exp(-\lambda)$ as

$$\exp(-\lambda) = [1 + U^2 - 8\pi(\rho+\varepsilon)\,R^2/3]\,(R')^{-2}, \tag{18}$$

where $U = u^a(\partial R/\partial x^a) = \exp(-\nu/2)\,\dot{R}$ denotes the invariant velocity of the fluid.

We now assume the motion of the fluid is shear free and hence from (12), we get

$$\dot{\lambda} = 2\dot{R}/R, \tag{19}$$

which on integration gives

$$\exp(\lambda/2) = R'/f(r), \tag{20}$$

where $f(r)$ arises out of integration and is related with the curvature of the 3-spaces t = constant.

The relation between λ and R given by (20) simplifies the metric of the space-time as

$$ds^2 = -(R')^2 d\bar{r}^2 - R^2(d\theta^2 + \sin^2\theta\, d\phi^2) + \exp\nu\, dt^2, \tag{21}$$

where we have written

$$d\bar{r} = dr/f(r). \tag{22}$$

Using (19), we simplify (16) as

$$\nu' = \dot{\lambda}'/\dot{\lambda}, \tag{23}$$

so that the whole system is reduced to a single variable $R(\bar{r}, t)$.

We now eliminate p from (13) and (14) and simplify (17) giving respectively

$$\partial^2(R^{-1})/\partial\bar{r}^2 = R^{-1} - B(\bar{r})R^{-2}, \tag{24}$$

and

$$\partial^2(R^{-1})/\partial\bar{r}^2 = R^{-1} - 4\pi\varepsilon R, \tag{25}$$

where $B(\bar{r})$ is a "constant" of integration. it was shown by Thompson and Whitrow (1967) that the necessary and sufficient condition for the density function to be uniform, viz. $\rho = \rho(t)$, is that $B(\bar{r})$ = constant.

Hence, by comparing (24) and (25), we get

$$\varepsilon = B/4\pi R^3, \tag{26}$$

which is a generalization of the result for the Schwarzschild interior solution showing that a field singularity at the centre is inevitable even in the present more general situation. Again we note both ρ and ε fall off as inverse cube of R.

4. Charged fluid spheres

The choice of spherical symmetry and comoving coordinates restricts the non-vanishing components of the Maxwell skew-symmetric tensor to just F_{14}.

Hence for charged perfect fluid spheres (13), (14) and (15) modify as below (Krishna Rao 1972b):

$$8\pi T_1^1 = -8\pi(p-K) = -[1/R\exp(\lambda)]\ \{(R'^2/R) + R'\nu'\}$$
$$+[1/R\exp(\nu)]\ \{(2\ddot{R} + (\dot{R}^2/R) - \dot{R}\dot{\nu}\} + (1/R^2), \tag{27}$$

$$8\pi T_2^2 = 8\pi T_3^3 = -8\pi(p+K) = 8\pi\varepsilon$$
$$-[1/R\exp(\lambda)]\ \{2R'' - (R'^2/R) - R'\lambda' + R'\nu'\}$$
$$+[1/R\exp(\nu)]\ \{2\ddot{R} - (\dot{R}^2/R) - \dot{R}\dot{\lambda} - \dot{R}\dot{\nu}\} - (1/R^2), \tag{28}$$

$$8\pi T_4^4 = 8\pi(\rho+K) = -[1/R\exp(\lambda)]\ \{2R'' + (R'^2/R) - R'\lambda'\}$$
$$+[1/R\exp(\nu)]\ \{(\dot{R}^2/R) + \dot{R}\dot{\lambda}\} + (1/R^2), \tag{29}$$

where

$$K = -(1/2)F_{14}F^{14} = |\bar{E}|^2 = q^2(r)/R^4. \tag{30}$$

In analogy with the classical electromagnetic case we call K the energy density of the electromagnetic field. With this choice of F_{14} Maxwell's equations are identically satisfied.

We once again compute the expression for $\exp(-\lambda)$ as

$$\exp(-\lambda) = [1 + U^2 - 8\pi(\rho + \varepsilon + 3K)R^2/3](R')^{-2}. \tag{31}$$

From (31) we note that both the energy density of the free gravitational field ε and the energy density of the electromagnetic field K are coupled to the material energy density ρ.

We now assume that the space-time is spatially isotropic so that (19)–(23) are satisfied. Hence once again, we can eliminate p from (27) and (30) using (19) and (23) so that

$$\partial^2(R^{-1})/\partial\bar{r}^2 = R^{-1} - B(\bar{r})R^{-2} + 8\pi KR, \tag{32}$$

where $B(\bar{r})$ is again a "constant" of integration. Similarly, the expression for ε can also be simplified to give the relation

$$\partial^2(R^{-1})/\partial\bar{r}^2 = R^{-1} - 4\pi\varepsilon R - 4\pi KR. \tag{33}$$

Comparing (32) and (33) we get

$$\varepsilon + 3K = B(\bar{r})/4\pi R^3. \tag{34}$$

Again, the condition for uniform density, $\rho = \rho(t)$, reduces $B(\bar{r})$ to a constant which when substituted in (31) gives a singularity in the metric tensor at the centre of the sphere. Thus, the combined energy density of the free gravitational field as well as the electromagnetic field produces a singularity at the centre of the sphere. The typical R^{-3} fall off for ρ and $(\varepsilon + 3K)$ shows the generic nature of the singularity.

5. Models filled with stiff fluid $p = \rho$

Defining the shear-expansion ratio ξ as (MacCallum, 1982),

$$\xi = \mid 3\sigma^{b}_{a}\sigma^{a}_{b} / 2\theta^{2} \mid^{1/2}, \tag{35}$$

and taking $\xi = 1/2$, we get (Krishna Rao, 1995)

$$\lambda = 0, \; R = r\, s(t), \; \exp \nu = r^{2}. \tag{36}$$

Making use of (36) in (13), (15) and (17), we obtain

$$8\pi p = 8\pi\rho = 4\pi\varepsilon = a^{2}/R^{2}. \tag{37}$$

The function $s(t)$ is given by the involved relation

$$(s^{4} - s^{2} + a^{2})^{1/2} + s^{2} - (1/2) = \exp(\pm 2t/t_{0}), \tag{38}$$

a and t_0 being constants of integration (see Wesson, 1978).

From (37) we note that both ρ and ε have the R^{-2} fall off typical for isothermal gas spheres (see Chandrasekhar, 1972). We also note that the field singularity dominates over matter singularity.

6. Concluding remarks

The Weyl curvature hypothesis (Penrose, 1977, 1979, 1981, 1986) makes a distinction between the Friedman-Robertson-Walker (as well as Tolman-Bondi) type of initial singularities where the Weyl tensor vanishes and the final (black hole) singularities with Weyl tensor diverging to infinity. At the "big bang" the matter with uniform density has maximum entropy ($\rho \to \infty$) whereas entropy of the gravitational field is zero ($\varepsilon = 0$). In view of the vanishing conformal Weyl tensor the geometry also has a regular conformal structure at the "big bang". On the other hand at the "big crunch" (the final cosmological or black hole singularities) the Weyl tensor dominates completely over the Ricci tensor raising the gravitational entropy to the maximum (i.e. ε diverges faster than ρ to infinity.)

However, the role of this gravitational entropy in relation to matter entropy during the final stages of collapse (near the singularity) is still not properly understood.

Similarly, the condition for a trapped surface

$$D_k R = \exp(-\nu/2)\, \partial R/\partial t + \exp(-\lambda/2)\, \partial R/\partial r < 0,$$

leads to (Krishna Rao and Annapurna, 1992)

$$4\pi R^2 > 3/2\ (\rho + \varepsilon), \tag{39}$$

showing that the free gravitational field energy density ε plays an important role in the formation of black holes and subsequent evolution of singularities.

References

1. H. Bondi (1947) Mon. Not. R. Astron. Soc. **107**, 410.
2. S. Chandrasekhar (1972) in General relativity, ed. L. O'Raifeartaigh, Clarendon Press, Oxford.
3. J. Krishna Rao (1966) Curr. Sci. **35**, 589.
4. J. Krishna Rao (1972a) J. Phys. (London), **A 5**, 479.
5. J. Krishna Rao (1972b) Proc. Matscience Conference, Bangalore.
6. J. Krishna Rao and M. Annapurna (1985) in A random walk in relativity and cosmology, ed. N. Dadhich et al, Wiley Eastern Ltd; New Delhi.
7. J. Krishna Rao and M. Annapurna (1992) Pramana - J. Phys. **38**, 21.
8. J. Krishna Rao (1995) J. Indian Math. Soc. **61**, 57.
9. M.A.H. MacCallum (1982) in The origin and evolution of galaxies, ed. V.de Sabbata, World Scientific Pub; Singapore.
10. R. Penrose (1969) Nuovo Cimento Soc. Ital. Fis. **1**, 252.
11. R. Penrose (1977) in Proc. of the first Marcel Grossman meeting on general relativity, ed. R. Ruffini, North-Holland Pub. Co. Amsterdam.
12. R. Penrose (1979) in General Relativity: An Einstein Centenary Survey, ed. S.W. Hawking and W. Israel, Cambridge University Press, Cambridge.
13. R. Penrose (1981) in Quantum gravity 2: A second Oxford symposium, ed. C. J. Isham, R. Penrose and D.W. Sciama, Clarendon Press, Oxford.
14. R. Penrose (1986) in Herman Weyl, ed. K. Chandrasekharan, Springer-Verlag, Berlin.
15. A.Z. Petrov (1954) Sci. Not. Kazan State Univ. **114**, 55.
16. F.A.E. Pirani (1957) Phys. Rev. **105**, 1089.
17. F.A.E. Pirani (1962a) in Recent developments in general relativity, Pergamon Press, New York.

18. F.A.E. Pirani (1962b) in Gravitation: An introduction to current research, ed. L. Witten, John Wiley & Sons Inc., New York.
19. I.H. Thompson and G.J. Whitrow (1967) Mon. Not. R. Astron. Soc. **136**, 207.
20. R.C. Tolman (1934a) Proc. Natl. Acad. Sci. (U.S.A.) **20**, 169.
21. R.C. Tolman (1934b) Relativity, thermodynamics and cosmology, Clarendon Press, Oxford.
22. R.C. Tolman (1949) Rev. Mod. Phys. **21**, 374.
23. P.C. Vaidya (1943) Curr. Sci. **12**, 183.
24. P.C. Vaidya (1951) Proc. Indian Acad. Sci. **A 33**, 264.
25. P.C. Vaidya (1952) Curr. Sci. **21**, 96.
26. P.C. Vaidya (1953) Nature, London, **171**, 260.
27. P.S. Wesson (1978) J. Math. Phys. **19**, 2283.

Inflation : The Current Trends

Kalyani Desikan
Cosmology Group, Department of Mathematics
Indian Institute of Technology
Madras – 600 036, India.

1. Introduction

Our current understanding of the evolution of the universe is based upon the Friedmann-Robertson-Walker (FRW) cosmological model, or the hot big bang model as it is usually called. The model is so successful that it has become known as the standard cosmological model.

In the standard model the universe is assumed to be homogeneous and isotropic and is described by the FRW metric

$$ds^2 = dt^2 - R^2(t)\left[\frac{dr^2}{1-kr^2} + r^2\,(d\theta^2 + \sin^2\theta\ d\phi^2)\right]$$

where $k = 0, \pm 1$. The evolution of the scale factor $R(t)$ is governed by

$$\frac{\ddot{R}}{R} = -\frac{4\,\pi\,G}{3}\,(\rho + 3p) \tag{1}$$

$$\left(\frac{\dot{R}}{R}\right)^2 + \frac{k}{R^2} = \frac{8\pi G\rho}{3} \tag{2}$$

The big bang model is successful in the following aspects:

i) it provides an adequate general scenario for the large scale features of the present-day universe e.g., expansion rate, overall isotropy etc.

ii) the primordial abundances of light chemical elements as calculated for the big bang model are in agreement with observation.

iii) it is consistent with the existence of the Cosmic Microwave Background Radiation (CMBR).

However, the successes of the standard big bang model depend critically on very special and unnatural assumptions about the initial conditions in the universe at earlier times. These unnatural assumptions give rise to the so-called horizon, flatness problems etc., and prevent the standard cosmology from providing an explanation for the formation of galaxies.

Section 2 deals with the problems inherent in the standard model. In section 3, inflation driven by symmetry breaking phase transitions i.e., old inflation, new inflation, chaotic inflation, extended inflation, etc., is presented. Inflation driven by bulk viscosity has been dealt with in section 4. In section 5, the thermodynamics of open systems has been discussed. This has then been applied to cosmology and it has been shown how particle creation out of gravitational energy can give rise to inflation.

2. Shortcomings of the Standard Model

(i) Large-scale smoothness or horizon problem

The most precise indication of the smoothness of the universe is provided by the CMBR, which is uniform to about a part in 10^4 on angular scales from 10 sec of arc to 180°. Were the universe very inhomogeneous or were the expansion anisotropic, comparable temperature anisotropies in the CMBR would exist.

If the entire observable universe were in causal contact when the radiation last scattered, microphysical processes could have smoothed out any temperature fluctuations, and a single temperature would have resulted. However, within the standard cosmology this could not have happened because of the existence of particle horizons. Hence, in standard cosmology there is no physical explanation for the large-scale smoothness of the universe.

(ii) Small-scale inhomogeneity

Although the universe is apparently very smooth on large scales, there is a plethora of structures on smaller scales; stars, galaxies, clusters, voids and superclusters. The existence of abundant structure in the universe today poses another puzzle for the standard cosmology. What is the origin of the primeval inhomogeneity? There is no explanation for this in the standard model.

(iii) Spatial-flatness or oldness problem

Consider the ratio of the critical energy density defined by

$$\rho_c = \frac{3H^2}{8\pi G}$$

to the actual density ρ and call this ratio Ω, the density parameter. From equation (2) we find

$$\Omega = 1 + \frac{k}{\dot{R}^2}$$

In the standard model since $\dot{R}$ decreases with time Ω always deviates further and further from 1 as time goes on. The present observational data restrict Ω to lie in the interval [0.01,]. The flatness problem is : after a span of 10 billion years of expansion of the universe, why Ω is still remarkably close to 1?

(iv) Unwanted relics

Within the context of unified gauge theories there are a variety of stable, superheavy particle species which should have been produced in the early universe and which could survive annihilation and contribute too generously to the present energy density i.e., $\Omega >> 1$. The standard cosmology has no mechanism of ridding the universe of relics that are overproduced early in the history of the universe.

(v) Entropy problem

This is a restatement of the horizon problem and flatness problem in a somewhat different form. Our universe is characterized by a considerable entropy content. The big bang model simply assumes that all matter and radiation come into existence instantly at the singular epoch $t = 0$. Thus, the model does not explain the origin of the CMBR; rather, it only accommodates its existence. This is due to the fact that the classical evolution equations of general relativity are purely adiabatic and reversible. As such they are unable to explain by themselves the origin of cosmological entropy which might have been generated through irreversible processes during the cosmic expansion.

(vi) Singularity problem

The primordial singularity is a consequence of extrapolation of the field equations of the standard model to the epoch $t = 0$. But there is no justification in extrapolating the field equations prior to the Planck epoch t_{pl}, as the classical field equations are valid only for $t > t_{pl}$.

The problems mentioned above cannot be explained by the standard model, but they can only be accommodated in the standard model by making unnatural assumptions about the initial conditions. In 1981 Alan Guth [1] proposed the so-called *inflationary phase* as the solution to these problems. The word "inflation" is supposed to indicate a rapid expansion.

The expansion of the universe would be rapid i.e., accelerating if $\ddot{R} > 0$. From (1) we see that this translates into a condition on the equation of state as

$$p < -\frac{1}{3}\rho.$$

Hence, it can be seen that negative pressure is required in order to obtain generalised inflation. Thus, we must go beyond the ideal gas description of matter to obtain inflation. There are various mechanisms to obtain inflation.

One possibility is to turn to a description of matter in terms of classical and quantum fields. Moreover, the standard model predicts that as we go backwards in time temperature increases. It is now well established that at high temperatures (or equivalently high energies) the classical description of matter as an ideal gas breaks down. Hence, we expect that an improved description of the early universe may be obtained by using the theory of matter for the primordial state of matter which holds at very high energies.

Exponential expansion arises for a wide range of cosmological models in which the matter source is temporarily dominated by a scalar field with some self-interaction potential $V(\phi)$. When the field enters a regime in which the evolution of ϕ is dominated by its potential energy then it exerts the same stress as an effective cosmological constant and this results in exponential expansion. However, if the potential energy of the ϕ – field is directly proportional to its kinetic energy then there will be power-law expansion i.e.,

$$R(t) \sim t^p,\ p > 1$$

rather than exponential expansion. This is known as "generalised" inflation [2,3].

In addition to these theories in which the inflationary expansion is driven basically by a scalar field displaced from the minimum of the potential, there are other mechanisms to drive inflation such as bulk viscosity, particle creation, R^2 gravity, inflation from extra dimensions etc.

We begin our discussion of "Inflation" with models in which symmetry-breaking phase transitions lead to inflation.

3. Inflation Driven by Symmetry-breaking Phase Transitions

(A) Old inflation

The old inflationary universe model is based on a scalar field theory which undergoes a first order phase transition. A double well potential as in fig.1. is postulated.

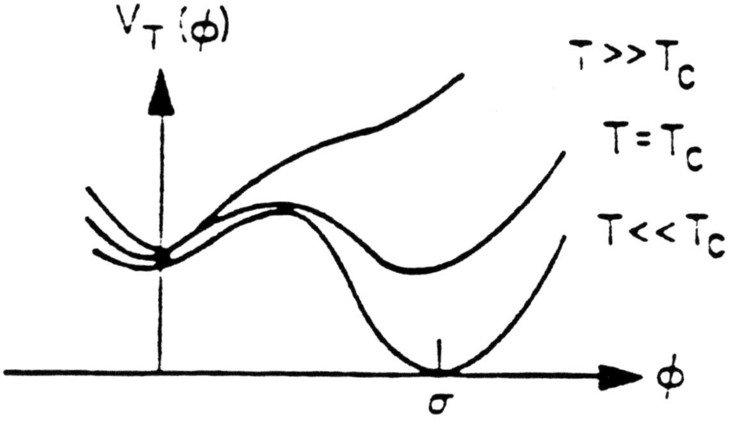

Fig.1

Here $\phi = 0$ is a local (metastable) minimum of $V(\phi)$ i.e., false vacuum and $\phi = \sigma$ is the global minimum i.e., true vacuum. In order to avoid a large cosmological constant at the present time, $V(\sigma) = 0$ is assumed. For $T > T_c$, the critical temperature, ϕ is confined to be close to zero by thermal forces. As the universe cools below T_c, ϕ gets stuck in the metastable false vacuum state which induces the exponential growth in the scale factor. The expansion rate $H \equiv \dot{R}/R$ is given by the FRW equation with $k = 0$ as

$$H^2 = \frac{8\pi G}{3} V(0)$$

Here inflation ends via the quantum-mechanical formation of bubbles of the true vacuum by tunneling. Such bubbles form with a characteristic size determined by microphysics. The bubbles then grow until they collide with the adjacent bubbles and this disperses the coherent energy in the bubble walls. But this scenario is flawed because the exponential expansion of the false-vacuum region generically dominates over bubble formation and so inflation never ends. This is known as the graceful exit problem. Moreover, the bubble walls, which carry a large fraction of the initial vacuum energy density, will remain inside our observed horizon. They would create energy-density perturbations of unacceptable magnitude. This problem stems from the fact that bubbles form after inflation.

(B) New inflation

The old inflation model was never considered to be a viable cosmological model because of the problems inherent in the model. To circumvent the problems in the old inflationary scenario, Linde [4] and Albrecht and Steinhardt [5] proposed a modified version of the scenario - the new inflationary universe. Since in this proposal bubbles form before inflation, the observed part of the universe lies within one large bubble and the above problems do not arise.

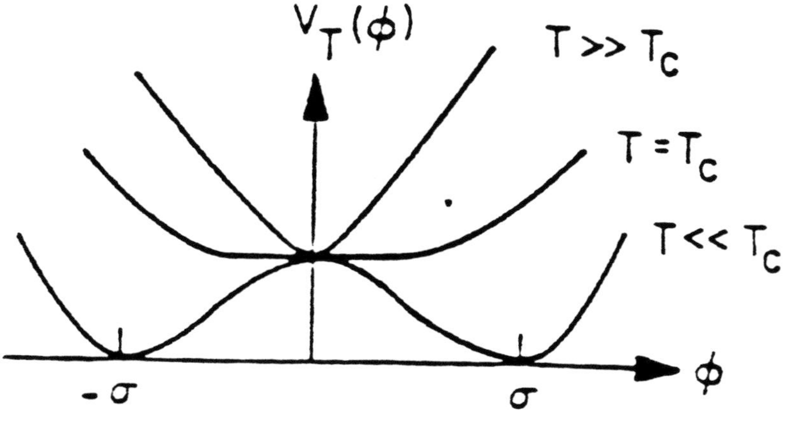

Fig.2

The starting point for this scenario is a scalar field theory with a double well potential, as in Coleman-Weinberg potential, which undergoes a second-order phase transition. Here $V(\phi)$ is symmetric and $\phi = 0$ is a local maximum of the zero temperature potential. Once again for $T \geq T_c$, ϕ is confined to values near $\phi = 0$ due to finite temperature effects. For $T < T_c$, thermal fluctuations trigger the instability of $\phi = 0$ and ϕ evolves towards $\phi = \pm\sigma$. Here there is no energy barrier separating the false vacuum $\phi = 0$ from the true vacuum $\phi = \pm\sigma$; instead the false vacuum lies at the top of a rather flat plateau at the time of phase transition. The scalar field ϕ will then start to move from near $\phi = 0$ towards either $\phi = \pm\sigma$. As long as ϕ remains close to zero, the energy density will be dominated by $V(\phi)$ and inflation proceeds. Eventually, ϕ will reach large values and nonlinear effects become important. ϕ finaly reaches $\pm\sigma$, overshoots and starts oscillating about the global minimum of $V(\phi)$. The amplitude of this oscillation is damped by the expansion of the universe and by coupling of ϕ to other fields. During the period of oscillation particles are produced which eventually equilibrate and regenerate a thermal bath. The reheating period will be very short and the universe will reheat to a temperature equal to T_c up to a factor of order unity. Thenceforth, cosmology evolves as in the standard model.

The new inflationary universe model — although it was for a long time presented as a viable model — has its share of problems as well. It suffers from severe fine tuning and initial condition problems. Since the whole concept of inflation was brought in to avoid fine tuning, this requirement is like breaking the ground rules.

C. *Chaotic Inflation*

The original model proposed by Guth involved a strongly first-order phase transition, while the new inflationary model required a weakly first-order or even a second-order phase transition. In the chaotic inflation model proposed by Linde [6] there is no phase transition involved. Chaotic inflation is based on the observation that for weakly coupled scalar fields, initial conditions which follow from classical considerations alone lead to very large values of ϕ. In this scenario at the initial time ϕ is very large, homogeneous and static. This initial value is believed to be due to chaotic initial conditions. The energy-momentum tensor will be dominated by the large potential energy term and will lead to an equation of state $p = -\rho$ giving rise to inflation. While one can produce sufficient inflation this way, it is necessary to ensure that the initial kinetic energy of the ϕ-field is small compared with the potential energy. Detailed calculations show that this requires the field to be uniform over sizes bigger than the Hubble radius.

D. *Extended inflation*

Extended inflation [7,8,9] is a revival of the spirit of Guth's old inflationary cosmology in the context of Brans-Dicke theory of gravity. The action is

$$S = \int dx^4 \sqrt{-g} \left[\phi R - \omega \frac{\partial_\mu \phi \, \partial^\mu \phi}{\phi} + 16\pi \mathscr{L}_m \right]$$

Here, ϕ is the Brans-Dicke scalar field, ω is the Brans-Dicke coupling parameter and $\mathscr{L}_m$ is the Lagrangian for matter fields. $\mathscr{L}_m$ is chosen to give rise to a first order phase transition which proceeds by nucleation of true vacuum bubbles in a surrounding sea of false vacuum.

The crucial point is that in Brans-Dicke theory, an equation of state $p = -\rho$ for matter leads to power-law rather than exponential expansion of the universe. For a FRW metric, the equations of motion are (for $k = 0$)

$$H^2 = \frac{8\pi\rho}{3\phi} + \frac{\omega}{6}\left(\frac{\dot{\phi}}{\phi}\right)^2 - H\frac{\dot{\phi}}{\phi}$$

and

$$\ddot{\phi} + 3H\dot{\phi} = \frac{8\pi(\rho - 3p)}{3 + 2\omega}$$

for $p = -\rho$, the solution is

$$R(t) = \left(1 + \frac{Ht}{\alpha}\right)^{\omega + 1/2}$$

and

$$\phi(t) = m_{pl}^2 \left(1 + \frac{Ht}{\alpha}\right)^2$$

with $\alpha \sim \omega^2$ and

$$H^2 = \frac{8\pi\rho}{3} m_{pl}^{-2}.$$

Since the false vacuum only expands as a power of t in Brans-Dicke theory, true vacuum bubbles are able to percolate i.e., the bubble nucleation rate always eventually overcomes the expansion and brings the inflationary era to a satisfactory end. Thus, the graceful exit problem of the old inflationary model can be overcome. It is noteworthy that the difficulties of old inflation can be circumvented in this way in any power-law or slower than exponential inflationary models.

However, percolation of true vacuum bubbles alone is not sufficient to give rise to a viable cosmological model. It was realised by Weinberg [10] and La, Steinhardt and Bertschinger [11], that the original extended inflation model fails, because the bubbles that nucleated early in inflation have time to grow to large sizes. The true vacuum within these large bubbles does not have time to thermalise before radiation decoupling and would create excessively large distortions in the microwave background. This conflict can be avoided if we place an upper bound on the BD parameter ω during inflation. According to Weinberg, thermalisation occurs only if $\omega < 10$, a bound which conflicts with the lower bound $\omega > 500$, at present, stemming from time delay measurements [12].

Hence, in order to save extended inflation, new variants of the model have been suggested. Several authors have considered extending the gravitational action beyond the BD action functional used in the original extended inflation model. The first method of evading the microwave bounds was to introduce a potential term $V(\phi)$ for the BD field [13]. This model, however, has the drawback that the BD potential must be very flat in order to avoid dominating the false-vacuum energy and thus disrupting the first-order inflation. Another alternative invokes different couplings of the BD field to visible and invisible matter [14] though these have recently been more strictly constrained [15]. One other possibility for successful extended inflation might be multidimensional theories such as Kaluza-Klein theories. After all, the major motivation for renewed interest in scalar-tensor theories such as Brans-Dicke theory is that an effective low-energy theory of the Brans-Dicke form follows naturally from superstring, supergravity and Kaluza-Klein theories. The most popular alternative is to make further modifications to the gravitational action leading to models referred to as *Hyper-extended Inflation* [16,17]. Another alternative would be to consider a more general scalar-tensor theory in which the BD coupling parameter ω evolves with time. Barrow and Maeda [18] have found examples of a new type of inflation intermediate between the power-law and exponential types. They assumed ω as a function of the scalar field ϕ which increases during the evolution of the universe. The functional dependence of ω was taken to be

$$\omega(\phi) = \omega_m \phi^m + \omega_0$$

where ω_m and ω_0 are dimensional positive constants and the power index m is a real number. Bellido [19] proposed a solution to extended inflation by choosing ω as

$$2\omega(\phi) + 3 = 2\beta\,(1 - \phi/\phi_c)^{-\alpha}$$

where $\phi_c = M_c^2$ with M_c a mass scale close to the Planck mass today, M_{pl}.

E. Inflation in Nordtvedt's theory

In Nordtvedt's theory, which is a general scalar-tensor theory, the coupling parameter $\omega = \omega(\phi)$ evolves with time. Here ω could be small initially ($\omega < 10$) and it could increase with the evolution of the universe to a large value ($\omega > 500$) at the present epoch. Hence, general scalar-tensor theories like Nordtvedt's theory, which overcome the problem inherent in working with the BD theory have caught the fascination of cosmologists [20].

The general scalar-tensor theory with $\omega = \omega(\phi)$ was originally proposed by Bergmann [21], Wagoner [22] and Nordtvedt [23]. The action in this case is

$$S = \int dx^4 \sqrt{-g}\left[\phi R - \omega(\phi)\,\frac{\partial_\mu \phi\, \partial^\mu \phi}{\phi} + 16\pi \mathscr{L}_m\right]$$

where ϕ is the scalar field, ω is the coupling parameter, being in general an arbitrary function of time and $\mathscr{L}_m$ is the matter Lagrangian.

The field equations corresponding to the above action, with the spatially flat ($k = 0$) FRW metric are

$$H^2 = \frac{8\pi\rho}{3\phi} + \frac{\omega}{6}\left(\frac{\dot\phi}{\phi}\right)^2 - H\frac{\dot\phi}{\phi}$$

and

$$\ddot\phi + 3H\dot\phi = \frac{8\pi(\rho - 3p)}{3 + 2\omega} - \frac{\omega'}{3 + 2\omega}\,\dot\phi^2.$$

The above equations lead to the continuity equation

$$\dot\rho + 3H(\rho + p) = 0.$$

Banerjee et al [20] have shown that for the equation of state $p = -\rho$, the old inflationary solution given in the form of an exponential function of time is fully consistent with a variable ω. Johri and Kalyani [24] have shown that for a barotropic equation of state

$$p = \gamma\rho, \quad 0 \le \gamma \le 1$$

and a constant deceleration parameter i.e.

$$q = \frac{-R\ddot{R}}{\dot{R}^2} = \text{constant}$$

it is possible to have both rapid power-law and exponential expansions, irrespective of the equation of state, with the coupling parameter ω increasing with time.

4. Inflation Driven by Bulk Viscosity

The role of bulk viscosity on the evolution of the universe has been investigated by several authors [25-32]. Bulk viscosity may arise under various circumstances. The FRW equations in the presence of bulk viscosity take the form

$$\left(\frac{\dot{R}}{R}\right)^2 + \frac{k}{R^2} = \frac{8\pi G\rho}{3} \tag{3}$$

$$\frac{\ddot{R}}{R} = -\frac{4\pi G}{3}(\rho + 3p - 3\mu\theta) \tag{4}$$

where μ is the co-efficient of bulk viscosity and $\theta \equiv 3\dot{R}/R$ is the expansion scalar. It can be noted that the effect of bulk viscosity is to reduce the thermodynamical pressure p by a factor $\mu\theta$; in this sense it plays the role of negative isotropic pressure, thereby decreasing the effect of self-gravitation and indirectly helping in the expansion of the universe. Bulk viscosity acts as a negative energy field and leads to creation of energy and entropy production in the universe. The functional form of μ would depend upon the specific nature of dissipation mechanism giving rise to bulk viscosity. But in a homogeneous model like Friedmann universe, μ will be a function of time only. Numerous investigations have been carried out on the effect of (i) constant bulk viscosity [25-27] and (ii) time dependent bulk viscosity on the evolution of FRW models [28-32].

From (4) it can be seen that bulk viscosity might lead to accelerated expansion and also to inflation if

$$3\mu\theta > \rho + 3p$$

It has been shown by Johri and Sudharsan [32] that a constant bulk viscosity however small, would ultimately drive an evolutionary flat Friedmann universe into a steady state asymptotically with the Hubble parameter, energy density etc., attaining stationary values despite the fact that the universe evolves.

5. Inflation Driven by Creation of Matter

As discussed in the Introduction, the standard big bang model is unable to explain the origin of cosmological entropy. This is due to the fact that the classical evolution equations of General Relativity are purely adiabatic and reversible. As such they are unable to explain by themselves the origin of cosmological entropy which might have been generated through irreversible processes during the cosmic expansion. Moreover, the classical field equations are valid only for $t > t_{pl}$; as such there is no justification in extrapolation of the field equations prior to Planck epoch t_{pl} and the primordial singularity might be non-existent [33,34,35]. As an alternative to the singular origin of the universe, the leading cosmologists [36] now favour a non-singular origin, the expansion starting from a vacuum fluctuation during the quantum era. In this context Prigogine's hypothesis of creation of matter out of gravitational energy is quite relevant.

A new type of cosmological history that includes large-scale entropy production was proposed by Prigogine et al [37,38,39]. The phenomenological macroscopic approach of Prigogine et al allows for both particle creation and entropy production. They have shown that the application of open thermodynamic systems to cosmology, leads very naturally to a reinterpretation of the matter stress-energy tensor in the Einstein's equations. Traditionally, the universe has been regarded as a closed system wherein the total number of particles remains constant. The universe could, however, be considered as an open system where the total number of particles is no more invariant with time. Particles could be continually created out of gravitational energy; this creation corresponds to irreversible energy flow from the gravitational field to the created matter constituents. In this scenario, the universe starts from a random vacuum fluctuation and the cosmic expansion is driven by the creation of matter particles.

In the standard treatment, the universe is considered as a closed system and the corresponding laws of thermodynamics have the form

$$d(\rho V) = dQ - p\,dV \tag{5}$$

and

$$T\,dS = d(\rho V) + p\,dV \tag{6}$$

where ρ is the energy density, p the thermodynamic pressure, V the volume, T the temperature and S the entropy of the system. It is seen from (5) and (6) that the entropy production is given by

$$T\,dS = dQ \tag{7}$$

Consequently, for a closed adiabatic systems ($dQ = 0$) the entropy remains stationary $dS = 0$. To account for the observed entropy in the observable universe, one has to either assume it as an initial condition or account for it through some dissipative mechanism. But, if one treats the universe as an open system, allowing for irreversible matter creation from the energy of the gravitational field, one can account for entropy production right from the beginning and also the second law of thermodynamics gets incorporated into the evolutionary equations in a more meaningful way.

Thus, allowing for matter creation, the appropriate analysis is performed in the context of open systems. In this case the number of particles N in a given volume V is not fixed to be a constant and the combined first and second law of thermodynamics reads

$$TdS = d(\rho V) + pdV - \mu dN \tag{8}$$

where μ is the chemical potential defined by the Euler's relation

$$\begin{aligned} \mu N &= (p + \rho) V - TS \\ &= \hat{H} - TS \end{aligned} \tag{9}$$

where $\hat{H} = (\rho + p) V$ is the enthalpy of the system. Using (9) in (8) leads to

$$TdS = d(\rho V) + pdV - (\hat{H} - TS)\, dN/N$$

This can be rewritten as

$$TS\,(dS/S - dN/N) = d(\rho V) + pdV - \hat{H}\, dN/N. \tag{10}$$

Now since the entropy $S = \sigma n V = \sigma N$ where σ is the specific entropy per particle and n is the particle number density we have

$$dS/S = d\sigma/\sigma + dN/N. \tag{11}$$

Using (11) in (10) we get the energy conservation equation for adiabatic open systems as

$$TN\, d\sigma = d(\rho V) + p\, dV - \hat{H}\, dN/N. \tag{12}$$

Equation (12) reduces to the energy conservation equation obtained by Prigogine et al when σ is a constant.

The second law of thermodynamics requires that $dS \geq 0$. We regard the second law of thermodynamics as one of the most fundamental laws of physics and it should hold whether creation of matter takes place or not. Therefore, from (11) we have

$$dN/N \geq -d\sigma/\sigma. \tag{13}$$

Hence, from (13) it can be seen that matter creation takes place if $d\sigma \leq 0$ and matter annihilation occurs if $d\sigma > 0$.

In fact, when $d\sigma = 0$, as in Prigogine's approach, (13) reduces to

$$dN \geq 0. \tag{14}$$

Therefore, in the particular case when σ is a constant, matter creation can take place, the reverse process (matter annihilation) being thermodynamically forbidden. This is the case considered by Prigogine et al. In this restricted case the energy conservation equation (12) can be written as

$$d(\rho V) + (p + p_c)\, dV = 0 \tag{15}$$

where

$$p_c = -\frac{\hat{H}}{N}\frac{dN}{dV} \tag{16}$$

p_c is negative or zero depending on the presence or absence of particle production. The energy conservation equation (15) suggests that, in the presence of particle creation the thermodynamic pressure is supplemented by an additional pressure p_c given by (16).

A) Modified Einstein's equations with creation of matter

In the presence of creation of matter (with $\sigma = 0$), the effective energy-momentum tensor of the cosmic fluid includes the creation pressure term p_c and is given by

$$\tilde{T}_{ab} = \left[\rho + p + p_c\right] u_a u_b - \left[p + p_c\right] g_{ab} \tag{17}$$

Accordingly the modified field equation is given by

$$G_{ab} = 8\pi G \tilde{T}_{ab} \tag{18}$$

Assuming the space-time to be given by the FRW metric

$$ds^2 = dt^2 - R^2(t)\left[\frac{dr^2}{1-kr^2} + r^2\left(d\theta^2 + \sin^2\theta\, d\phi^2\right)\right]$$

where $R(t)$ is the scale factor, the field equations (18) with the above metric and the barotropic equation of state

$$p = \gamma \rho, \qquad 0 \leq \gamma \leq 1 \tag{19}$$

now yield

$$\left(\frac{\dot{R}}{R}\right)^2 + \frac{k}{R^2} = \frac{8\pi G \rho}{3} \tag{20}$$

$$2\frac{\ddot{R}}{R} + \left(\frac{\dot{R}}{R}\right)^2 + \frac{k}{R^2} = -\,8\pi G\,(p + p_c) \tag{21}$$

Equations (20) and (21) lead to the continuity equation,

$$\dot{\rho} + 3H\,(\rho + p + p_c) = 0$$

or

$$\dot{\rho} + 3\,(1 + \gamma)\,\rho H = -\,3p_c H \tag{22}$$

where

$$p_c = -\,(1 + \gamma)\,(\rho / N)\,\frac{dN}{dt}\,(1/3H) \tag{23}$$

and $H = \dfrac{\dot{R}}{R}$ is the Hubble function.

In the particle creation scenario, the entire behavior of the model including its expansion, deceleration, entropy content etc., depend essentially upon the functional form of $N(t)$ i.e., the total number of particles as a function of time, hereafter called particle creation function. In fact, the different regions of the universe might evolve with different forms of $N(t)$ due to random vacuum fluctuations. Therefore, it would be quite logical to regard the particle creation function $N(t)$ as an initial condition. This makes the system of field equations (20) and (21) well defined and solvable.

For a flat ($k = 0$) universe models with a rate of creation of the form,

$$\frac{\dot{N}}{N} = aH, \quad a = \text{constant} \tag{24}$$

and

$$0 < \frac{(1 + 3\gamma)}{(1 + \gamma)} < a \leq 3$$

the expansion driven by creation would be accelerating.

For values of 'a' in the range

$$0 < \frac{(1+3\gamma)}{(1+\gamma)} < a < 3$$

there would be power-law inflation, whereas, $a = 3$ corresponds to deSitter (exponential) expansion

$$R(t) = R_0\, e^{\chi t} \tag{25}$$

Now using (25) in (20) with $k = 0$ leads to

$$\rho = \rho_0 = (3\chi^2/8\pi G) = \text{constant} \tag{26}$$

From equations (26) and (23) we get

$$p_c = -(1+\gamma)\,\rho = -(1+\gamma)\,\rho_0$$

Also the number of particles, entropy and particle number density are given by

$$N = N_0 R^3 = N_0 \exp 3\chi t$$

$$S = S_0 R^3 = S_0 \exp 3\chi t$$

$$n = N/R^3 = N_0 = \text{constant}$$

Hence, when $a = 3$, the energy density, thermodynamic pressure, creation pressure and particle number density remain constant, while the number of particles and entropy increase exponentially in this model. As such this model represents a steady-state inflationary universe.

B) Modified Brans-Dicke field equations with creation of matter

The modified field equations of Brans-Dicke (BD) theory are given by

$$G_{ab} = -\frac{8\pi}{\phi}\tilde{T}_{ab} - \frac{\omega}{\phi^2}\left[\phi_{;a}\,\phi_{;b} - \frac{1}{2}g_{ab}\,\phi_{;c}\,\phi^{;c}\right] - \frac{1}{\phi}\left[\phi_{;a;b} - g_{ab}\,\Box^2\phi\right] \tag{27}$$

$$\Box^2\phi = \frac{8\pi}{3+2\omega}\tilde{T}^{a}_{\cdot a} \tag{28}$$

Here $\tilde{T}_{ab}$ is the effective energy-momentum tensor of the cosmic fluid in the presence of creation of matter given by (17).

The BD field equations (27–28) with the FRW metric and barotropic equation of state reduce to

$$3\left(\frac{\dot{R}}{R}\right)^2 + 3\frac{k}{R^2} + 3\frac{\dot{R}}{R}\frac{\dot{\phi}}{\phi} - \frac{\omega}{2}\left(\frac{\dot{\phi}}{\phi}\right)^2 = \frac{8\pi\rho}{\phi} \tag{29}$$

$$2\frac{\ddot{R}}{R} + \left(\frac{\dot{R}}{R}\right)^2 + \frac{\ddot{\phi}}{\phi} + \frac{\omega}{2}\left(\frac{\dot{\phi}}{\phi}\right)^2 + 2\frac{\dot{R}}{R}\frac{\dot{\phi}}{\phi} + \frac{k}{R^2} = -\frac{8\pi\gamma\rho}{\phi} - \frac{8\pi p_c}{\phi} \tag{30}$$

$$\ddot{\phi} + 3\frac{\dot{R}}{R}\dot{\phi} = \frac{8\pi}{3+2\omega}\left[(1-3\gamma)\rho - 3p_c\right] \tag{31}$$

Equations (29)–(31) lead to the continuity equation given by (22).

As mentioned in the previous subsection $N(t)$ may be regarded as an initial condition in such models. Hence, by choosing the particle creation function $N(t)$, *ab* initio, the system of equations (29)–(31) would become well-defined and unique solutions can be sought.

For $k = 0$ models, a rate of creation of the form

$$\frac{\dot{N}}{N} = aH, \qquad a = \text{constant}$$

with 'a' lying in the range

$$3 + \frac{[-4 + 2(1 + 3/2\omega)^{1/2}]}{(1+\gamma)} < a \leq 3 + \frac{[-3 + 3(1 + 4/3\omega)^{1/2}]}{(1+\gamma)}$$

would lead to accelerated expansion.

Now $a = 3$ leads to an inflationary model with the scale factor, the BD scalar field and the energy density being given by

$$R(t) = (1+t)^{\omega + 1/2}$$

$$\phi(t) = K(1+t)^2$$

$$\rho(t) = \rho_0$$

$$p_c = -(1+\gamma)\rho_0$$

Since in this case

$$\dot{S}/S = \dot{N}/N = 3\,\dot{R}/R$$

we get

$$N(t) = N_0 (1 + t)^{3(\omega + 1/2)}$$

$$S(t) = S_0 (1 + t)^{3(\omega + 1/2)}$$

$$n \equiv N/R^3 = N_0, \text{ a constant}$$

Hence, the number of particles and the entropy increase indefinitely in this model, while the energy density, thermodynamic pressure, creation pressure and particle number density remain stationary. It furnishes another example of an inflationary steady-state model.

The inflationary models in BD theory as proposed by Mathiazhagan and Johri [7] and La and Steinhardt [8,9] are power-law models. Finally we present an exponentially expanding model in BD theory.

For $a = 3 \dfrac{[\gamma + (1 + 4/3\omega)^{1/2}]}{(1 + \gamma)}$ the scale factor grows exponentially i.e.,

$$R(t) \sim \exp(\chi t)$$

with the BD scalar field, energy density and creation pressure given by

$$\phi(t) = K \exp([-3 + 3(1 + 4/3\omega)^{1/2}]\, \chi t)$$

$$\rho(t) = \rho_0 \exp([-3 + 3(1 + 4/3\omega)^{1/2}]\, \chi t)$$

$$p_c = -\rho_0 [\gamma + (1 + 4/3\omega)^{1/2}] \exp([-3 + 3(1 + 4/3\omega)^{1/2}]\, \chi t)$$

where

$$\rho_0 = (3K/8\pi) [3(1 + \omega)(1 + 4/3\omega)^{1/2} - (4 + 3\omega)]\, \chi^2$$

Again since

$$\dot{S}/S = \dot{N}/N = a\chi$$

we have

$$N(t) = N_0 \exp(a\chi t)$$

$$S(t) = S_0 \exp(a\chi t)$$

$$n = N_0 \exp[(a - 3)\, \chi t]$$

Here ρ, ϕ, p_c, N, S and n also increase exponentially.

6. Conclusions

Inflation has been fairly successful in overcoming most of the problems plaguing the standard big bang model and there are various mechanisms that can lead to inflation. But inflation has not been able to provide an answer to the origin of structures; the spectrum of density perturbations that arise in these inflationary models would not give rise to the structures that we observe around us.

Acknowledgement

This work was done with financial support from Department of Science and Technology (DST), New Delhi under the project DST/92-93/007.

References

1. Guth, A.H., (1981), *Phys. Rev.*, **D23**, 347.
2. Lucchin, F. and Mataresse, (1985), *Phys. Rev.*, **D32**, 1316.
3. Abbott, L.F. and Wise, M.B., (1984), *Nucl. Phys.*, **B244**, 541.
4. Linde, A., (1982), *Phys. Letts.*, **B108**, 389.
5. Albercht, A. and Steinhardt, P.J., (1982), *Phys. Rev Letts.*, **48**, 1220.
6. Linde, A., (1983), *Phys. Letts.*, **B129**, 177.
7. Mathiazhagan, C. and Johri, V.B., (1984), *Class. Quantum Grav.*, **1**, L29.
8. La, D. and Steinhardt, P.J., (1989), *Phys. Rev Letts.*, **62**, 376.
9. La, D. and Steinhardt, P.J., (1989), *Phys. Letts.*, **B220**, 375.
10. Weinberg, E.J., (1989), *Phys. Rev.*, **D40**, 3950.
11. La, D., Steinhardt, P.J. and Bertschinger, E.W., (1989), *Phys. Letts.*, **B231**, 231.
12. Reasenberg et al., (1979), *Astrophys. J.*, **234**, L219.
13. Accetta, F.S. and Trester, J.J., (1989), *Phys. Rev.*, **D39**, 2854.
14. Holman, R., Kolb, E.W. and Wang, Y., (1990), *Phys. Rev. Letts.*, **65**, 17.
15. Damour, T. and Gundlach, C., (1991), *Phys. Rev.*, **D43**, 3873.
16. Steinhardt, P.J. and Accetta, F.S., (1990), *Phys. Rev. Letts.*, **64**, 2740.
17. Little, A.R. and Wands, D., (1992), *Phys. Rev.*, **D45**, 2665.
18. Barrow, J.D., Maeda, K., (1990), *Nucl. Phys.*, **B341**, 294.
19. Bellido, J.G. and Quiros, M., (1990), *Phys. Letts.*, **B243**, 45.
20. Banerjee, A., Choudhury, S.B.D., Banerjee, N. and Sil, A., (1993), *Pramana*, **40**, 31.
21. Bergmann, P.G., (1968), *Int. J. Theor. Phys.*, **1**, 25.
22. Wagoner, R.V., (1970), *Phys. Rev.*, **D1**, 3209.

23. Nordtvedt, K., (1970), *Astrophys. J.*, **161**, 1059.
24. Johri, V.B. and Kalyani Desikan, (1994) *Pramana-Jl. of Physics*, **42** (6), 473.
25. Heller, M. and Klimek, Z., (1973), *Astrophys. Space Sci.*, **20**, 205.
26. Heller, M. and Suszychi, L., (1974), *Acta. Phys. Pol.*, **B5**, 345.
27. Padmanabhan, T. and Chitre, S.M., (1987), *Phys. Lett.*, **A120**, 433.
28. Nightingale, J.D., (1973), *Astrophys. J.*, **185**, 105.
29. Santo, N. and Banerjee, A., (1985), *J. Math. Phys.*, **26**, 878.
30. Waga, I., Falcao, R.C. and Chanda, R., (1986), *Phys. Rev.*, **D33**, 1839.
31. Pacher, T., Stein-Schabes, J.A. and Turner, M.S., (1987), *Phys. Rev.*, **D36**, 1603.
32. Johri, V.B. and Sudharsan, R., (1988), *Phys. Lett.*, **A132**, 316.
33. Hawking, S.W., (1979), *In General Relativity : An Einstein Centenary Survey*, eds., S.W. Hawking and W. Tsrael, Cambridge University Press, Cambridge.
34. Hawking, S.W. and Halliwell, J., (1985), *Phys. Rev.*, **D31**, 1777.
35. Narlikar, J.V. and Padmanabhan, T., (1986), *Gravity, Gauge Theories and Quantum Cosmology*, Reidel, Dordrecht.
36. Hawking, S.W., (1988), *A Brief History of Time*, Bantam Book, **141**.
37. Prigogine, I., Geheniau, J., (1986), *Proc. Nat. Acad. Sci., U.S.A.*, **83**, 6245.
38. Prigogine, I., Geheniau, J., Gunzig, E. and Nardone, P., (1988), *Proc. Nat. Acad. Sci., U.S.A.*, **85**, 7428.
39. Prigogine, I., Geheniau, J., Gunzig, E. and Nardone, P., (1989), *Gen. Relativ. Grav.*, **21**, 767.

Torsion and Quantum Gravity

Venzo de Sabbata
World Laboratory, Lausanne, Switzerland
Dipartimento di fisica dell 'Universitá di Bologna e Ferrara
INFN-Istituto Nazionale di Fisica Nucleare, Sezione di Ferrara Italy

1. Introduction

This is a symposium on Early Universe and I would like to speak on Torsion and Quantum Gravity because, first of all, when one tries to investigate what happens in the early universe, one has to look into elementary particle physics and then one must take into account two original properties of every elementary particle that is mass and spin.

Einstein's theory of gravity, that is General Relativity, takes into account only the mass. This is good for macroscopic body: the mass is the source of gravity in the sense that the mass is responsible for the curvature of space-time. We also know that the general theory of relativity is the best and the simplest theory of gravitation which is in agreement with all experimental facts in the domain of macrophysics including the more recent experiments on time-delay, with radar on Mercury and Venus and other sophisticated experiments in the solar system.

But when we consider the early universe, we know that the cosmological problem is strictly connected with elementary particle physics. But when we consider general relativity together with elementary particle physics, the latter being described by quantum field theory, we are obliged to take into account not only the mass of elementary particles, but also the spin. In fact, the elementary particles are characterized not only by mass but also by the spin which occurs in units of $\hbar/2$. Mass and spin are two elementary and independent original concepts. Mass distribution in a space-time is described by energy-mementum tensor; similarly spin distribution is described in a field theory by a spin density tensor. While mass is connected with the curvature of space-time, spin will be connected with another geometrical property of the space-time, consequently, we must modify the general theory of relativity in order to connect this new geometrical property with the spin density tensor. In this way we are led to the notion of torsion. In fact, all elementary particles can be classified by means of the irreducible unitary representation of the Poincaré group and can be labelled with the translational part of the Poincaré group, while spin is connected with the rotational part. In a classical field theory, mass corresponds to a canonical stress- energy-momentum tensor and spin to a canonical spin tensor. The dynamical relation between the stress-energy-momentum tensor and curvature is expressed in general relativity by Einstein equations, hence there must be an analogous dynamical relation including spin density tensor. Since this is impossible in the framework of general relativity we are forced to introduce this new geometrical property that we call torsion. We can say that just as the mass is responsible for curvature, spin is responsible for torsion. We will now see from a formal point of view as to how we must modify general relativity theory and how it represents a slight modification of this theory. In fact, the main idea is to assume an affine connection asymmetric instead of the symmetric connection that we have in Einstein theory (the Christoffel symbols). Torsion is, in fact, connected with the antisymmetric part of the affine connection as we shall see.

The motivations are mainly of theoretical character. In fact we are, at first sight, in the presence of two very different worlds. On the one hand we have at a microscopical level the strong and weak interactions while the gravitational interaction is the weakest and does not seem to play any role at that level; on the other hand all known interactions, except gravitation, that is strong, weak and electromagnetic are well described within the framework of relativistic quantum field theory in flat Minkowski space-time. So at first sight it seems that gravitation has no effects when we are dealing with elementary particle physics. But today we know that this is not true. In fact, if we consider the quantum theory in curved space-time instead of in flat Minkowski space-time, we have some very important new effects (neutron interferometry, spin-spin contact interactions, non-linear spinor effects etc.). Moreover, at a microphysical level, that is when we are dealing with elementary particle physics the role of gravitation becomes very important and necessary, for instance, when we deal with the early universe.

Moreover, if we consider the early universe, this is the place where we must consider and apply both general relativity and quantum theory. But general relativity is a classical field theory and so we faced with the problem of quantizing gravity also if we want to be coherent and consistent in our approach to the physics of early universe.

To sum up: when we are dealing with a microphysical realm we find that, besides the mass, the spin comes into play and then it has to be considered as the source of a gravitational field (as the mass): remember that when we speak of a gravitational field, we mean a field inseparably coupled to the geometry, that is, we speak of the structure of the space-time. In analogy to the mass where the energy-momentum tensor is coupled to the metric, we will expect the spin-density tensor to be coupled to some geometrical quantity of the space-time (a quantity that should relate to the rotational degrees of freedom in space-time).

In this way we are led to a generalization of a Riemann space-time. This generalization was proposed in 1922 [1] by Cartan. He related the torsion tensor to the density of intrinsic angular momentum well before the introduction of the modern concept of spin. According to Trautman "the Einstein-Cartan theory is the simplest and the most natural modification of the original Einstein theory of gravitation" [2].

From the geometrical point of view, torsion is simply the antisymmetric part of an asymmetric affine connection $\Gamma^{\mu}_{\alpha\beta}$ that is

$$Q^{\mu}_{\alpha\beta} = (1/2)\left(\Gamma^{\mu}_{\alpha\beta} - \Gamma^{\mu}_{\alpha\beta}\right) \equiv \Gamma^{\mu}_{[\alpha\beta]} \tag{1}$$

and has a tensor character. In the presence of torsion the space-time is called Riemann-Cartan manifold and is denoted by U_4 (the Riemann space is denoted by V_4).

We will not go through the Einstein-Cartan theory as it follows very nearly the structure of general relativity. But we will emphasize the fundamental thing that one of the most important geometrical properties of torsion is that a closed contour in an U_4 manifold becomes, in general, a non-closed contour in the flat space-time V_4. This non-closure property, that is the fact that the integral

$$l^{\alpha} = \oint Q^{\alpha}_{\beta\gamma}\, dA^{\beta\gamma} \neq 0 \tag{2}$$

over a closed infinitesimal contour is different from zero, can be treated as defects in space-time in analogy to the geometrical description of dislocations (defects) in crystals: and this can constitute a way towards the quantization of gravity, that means, quantization of the space-time itself and this will be the main argument of this lecture.

2. Quantum gravity and torsion

One thing that I would like to say is that torsion may constitute a way towards quantization of gravity. In fact, we will show that by introducing torsion in General Relativity, that is by considering the effect of the spin and linking the torsion to defects in space-time topology, we can have a minimal unit of length and minimal unit of time.

We know that one of the major difficulties in contemporary physics, both in quantum field theory and in general relativity, is the presence of singularities and of infinities. In general, this depends on the fact that we are used to considering point particles and hence we face the divergence of self energy integrals that go to infinity, or in general relativity when we have to deal with a collapsing body (i.e. a black hole), we face singularities.

These difficulties may disappear if, with torsion, we have a minimal unit of length and time: that is if we consider a discretized space-time. In fact, if we want to quantize gravity we cannot follow the procedure of quantization that we normally use to quantize, for instance, other forces like weak, strong or electromagnetic. In fact, gravity is not a force but is the curvature (and torsion!) of the space-time and we must try to quantize the space-time itself.

Now the torsion gives rise to defects in space-time topology. We know that in the geometrical description of crystal dislocations and defects, torsion plays the role of defect density and in this context one can consider space-time as an elastic deformable medium in the sense of Sakharov. But before proceeding further, we would like to show that the absence of torsion leads to some difficulties.

In fact, in a recent paper Treder and Borzeskowski [3] have pointed out that the Landau-Peierls type of uncertainty relations

$$\Delta F(c\Delta t)^2 \geq (\hbar c)^{1/2} \tag{3}$$

(F denotes the magnitude of the field strength F_{ik} and ΔF is measurement inaccuracy, Δt is the time interval between two field measurements) could correspond to commutation rules of the form:

$$i[F_{ik}, dx^i \wedge dx^k] = \sqrt{\hbar c}\,, \tag{4}$$

In this case inequality (4) would have the meaning of quantum uncertainty relations, that is it would imply, via Rosenfeld's analysis, the uncertainty inequalities for quantum general relativity as:

$$\Delta g_{ik}\,\Delta x^i\,\Delta x^k \geq \hbar G/c^3 \tag{5}$$

and

$$\Delta\Gamma_{ikl}\,\Delta x^i\,\Delta x^k\,\Delta x^l \geq \hbar G/c^3 \tag{6}$$

(Δg and $\Delta\Gamma$ are the corresponding inaccuracies). However, they note that in general relativity the "field strength" Γ^i_{kl} (which are the usual Christoffel symbols) are not tensorial quantities. So distances cannot be defined independent of the field quantities g_{ik}.

In fact, as demonstrated in [3], one should not expect to find commutation rules corresponding to Rosenfeld's inequality relations

$$g_{ik} L_0^2 \geq \hbar G/c^3$$

$$\Gamma^i_{kl} L_0^3 \geq \hbar G/c^3 \qquad (7)$$

(L_0 denotes the dimension of the spatial region over which the average value of g_{ik} and Γ^i_{kl} is measured) because the quantities, "field strength" and "length", appearing in (7) cannot be defined independent of each other. The independent definition of those quantities for which commutation rules are to be formed is however necessary. Therefore, the inequalities (7) do not have the status of uncertainty relations like (5) and (6); they are rather relations describing the inaccuracy of the measurement of the single quantities g_{ik}, Γ^i_{kl}, etc. As a consequence of the dependence of distances on the metric g_{ik}, the inequalities (7) say something about the limitations on distance measurements. Indeed, a coordinate system in which $|g_{00}| = 1$, with the condition $|\Delta g_{00}| < |g_{00}|$ one is led from the first of (7) to

$$L_0^2 > |\Delta g_{00}| \, L_0^2 \geq \hbar G/c^3 \qquad (8)$$

showing that for distances of the order of magnitude of Planck's length the notion "length" loses it physical meaning. But this statement about limitations on field and length measurements does not coincide with the statement implied by uncertainty relations of type (3) and valid for canonically conjugate quantities.

Therefore, the conclusion drawn in Ref.[3] was that the inequalities (7) establish limitations on the validity of the relativistic quantum field conception. And it was stressed that such limitations should only occur in quantized General Relativity Theory but not in any other theory of quantum gravity.

This problem may not arise if torsion is considered, since in this case the asymmetric part of the connection i.e. $\Gamma^{\alpha}_{[\beta\gamma]}$, i.e. the torsion tensor $Q^{\alpha}_{\beta\gamma}$ is a true tensorial quantity. In fact, as we have said, with torsion one can define distances in the following way. Suppose we consider a small closed circuit and write $l^{\alpha} = \oint Q^{\alpha}_{\beta\gamma} dA^{\beta\gamma}$, where $dA^{\beta\gamma} = dx^{\beta} \wedge dx^{\gamma}$ is the area element enclosed by the loop, then l^{α} represents the so called "closure failure", i.e. torsion has a intrinsic geometric meaning that it represents the failure of the loop to close, l^{α} having the dimension of length. ($Q^{\alpha}_{\beta\gamma}$ has dimension of inverse length, dA is area).

In the above relation, we know that torsion can be related to the intrinsic spin $\hbar$ and as the spin is quantized, we can say that the defect in space-time topology should occur in multiples of the Planck length, $(\hbar G/c^3)^{1/2}$ i.e.

$$\oint Q^{\alpha}_{\beta\gamma}\, dx^{\beta} \wedge dx^{\gamma} = n\,(\hbar G/c^3)^{1/2}\, n^{\alpha} \tag{9}$$

(n is an integer and n^{α} = unitpointvector).

This is analogous to the well known $\oint pdq = n\hbar$, i.e. the Bohr-Sommerfeld relation. As $Q^{\alpha}_{\beta\gamma}$ plays the role of the field strength (analogous to F_{ik} for electromagnetism), relation (9) is analogous to that of eq.(4) (with $\sqrt{\hbar}$ on the right hand side). So distance has been defined independently of g_{ik}. In fact (9) would define a minimal fundamental length, i.e. the Planck length entering through the minimal unit of spin or action $\hbar$. So $\hbar$ deals with the intrinsic defect built into torsion structure of space time through [4] $l^{\alpha} = \oint Q^{\alpha}_{\beta\gamma}\, dx^{\beta} \wedge dx^{\gamma}$, i.e. we have

$$[Q^{\alpha}_{\beta\gamma}, dx^{\beta} \wedge dx^{\gamma}] \geq (\hbar G/c^3)^{1/2}\, n^{\alpha} \tag{10}$$

i.e. $\hbar$ is related to a quantized timelike vector with dimensions of length related to an intrinsic geometrical structure with torsion (after all we know that intrinsic spin $\hbar$ is a universal attribute of all matter). So torsion must enter into geometry.

In conclusion [5] one has a classical background metric $g_{\alpha\beta}$ (specifying $dx^{\beta} \wedge dx^{\gamma}$) and an (nearly) independent tensor field $Q^{\alpha}_{\beta\gamma}$ quantized via condition (9):

$$\oint Q^{\alpha}_{\beta\gamma}\, dx^{\beta} \wedge dx^{\gamma} = n(\hbar G/c^3)^{1/2}\, n^{\alpha} \tag{11}$$

This quantization leads to the commutation rules (10):

$$[Q^{\alpha}_{\beta\gamma}, dx^{\beta} \wedge dx^{\gamma}] = n(\hbar G/c^3)^{1/2}\, n^{\alpha} \tag{12}$$

and to corresponding uncertainty relations:

$$\Delta Q^{\alpha}_{\beta\gamma} (\Delta x^{\beta} \wedge \Delta x^{\gamma}) = (\hbar G/c^3)^{1/2}\, n^{\alpha} \tag{13}$$

Therefore, the Einstein-Cartan theory of gravitation should in contrast to Einstein's General Relativity Theory provide genuine quantum-gravity effects.

We can also observe from relation (9), that by considering the fourth component, we can define time in the quantum geometric level through torsion as

$$t = (1/c) \oint Q\, dA = n\,(\hbar G/c^5)^{1/2} \tag{14}$$

So torsion is essential to have a minimum unit of time $\neq 0$!

This in fact gives us the smallest definable unit of time as $(\hbar G/c^5)^{1/2} \cong 10^{-43}s$. In the limit as $\hbar \Rightarrow 0$ (classical geometry of general relativity) or $c \Rightarrow \infty$ (Newtonian case), we would recover the unphysical $t \Rightarrow 0$ of classical cosmology or physics. So both $\hbar$ and c must be finite to give a geometric unit for time (i.e $\hbar \Rightarrow 0$ and $c \Rightarrow \infty$ are equivalent). The fact that $\hbar$ is related to a quantized timelike vector, discretizes time. This quantum of time or minimal unit of time correspondingly implies a limiting frequency of $f_{\max} \approx (c^5/\hbar G)^{1/2}$. This would have consequences even for perturbative QED, in estimating self energies of electrons and other particles, i.e. the self energy integral (in momentum space) taken over the momenta of all virtual photons. To make the integral converge Feynman, in his paper on QED [6], multiplied the photon propagator, k^{-2}, by the ad hoc factor $-f^2/(k^2-f^2)$, where k is the frequency (momentum) of the virtual photon. This convergence factor, although it preserves relativistic invariance, is objectionable because of its ad hoc character without any theoretical justification. Feynman considers f to be arbitrarily large without definite theoretical basis. Here the presence of space-time defects associated with the torsion due to the intrinsic spin would give a natural basis for the maximal value for $f^2_{\max}$ as (from eq.(14))

$$\approx c^5/G\hbar \approx 10^{86} \tag{15}$$

(and extremely large as required Feynman), giving finite result (instead of ∞) for the self energy. This makes $f_{\max}$ another fundamental constant for particle physics serving as a high frequency cut off which is not arbitrary.

But we can go a little further and see if there are conjugate variables for which we can write commutation relations. We will see that R and Q can be such conjugate variables. Consider, in fact, the equation of geodesic deviation of a free particle with non-zero spin (in which coupling of spin to the space-time curvature occurs):

$$\frac{d^2x^\mu}{ds^2} + \begin{Bmatrix} \mu \\ \alpha\beta \end{Bmatrix} \frac{dx^\alpha}{ds}\frac{dx^\beta}{ds} = -2Q^\mu_{\alpha\beta}\frac{dx^\alpha}{ds}\frac{dx^\beta}{ds} \tag{16}$$

Following the above discussion, in the quantum picture the right hand side is proportional to multiples of the Planck length. As the left and side of eq.(16) is obtained from variation of $\int g_{\alpha\beta}\,(dx^\alpha/ds)\,(dx^\beta/ds)$, we would simply have

$$\Delta\Gamma^\mu_{\alpha\beta}\, dx^\alpha \wedge dx^\beta = \left[Q^\mu_{\alpha\beta},\, dx^\alpha \wedge dx^\beta\right] \geq (\hbar G/c^3)^{1/2}\, n^\mu \tag{17}$$

In the limit of vanishing torsion, we have the usual geodesic equation. So torsion in eq.(16) can be identified with the quantum correction (in multiples of $\hbar$) to the classical equation of motion.

Unlike electromagnetism where a single particle suffices to measure the components of the field tensor, in the case of gravitation we require the relative acceleration between two particles to specify the components of the curvature tensor, i.e. $R^{\alpha}_{\beta\gamma\delta}$ is defined through the acceleration:

$$d^2x^{\alpha}/ds^2 = R^{\alpha}_{\beta\gamma\delta}\,(dx^{\beta}/ds)\,(dx^{\gamma}/ds)\,k^{\delta} \tag{18}$$

(k^{δ} = unit vector in the direction of the line joining the two particles). As many approaches lead to the notion of "maximal acceleration", for instance in the case of gravitation, this is of $\sim m_{Pl}\,c^3/\hbar = c^{7/2}/(\hbar G)^{1/2}$ (m_{Pl} = Planck mass), then the maximal value for d^2x^{α}/ds^2 would also imply a maximal curvature which turns out to be $\approx c^3/\hbar G \approx 10^{66} cm^{-2}$. In the classical limit, i.e. $\hbar \Rightarrow 0$, we have the usual curvature singularity, i.e. the maximal value is infinite, the relation between acceleration and curvature being of the form $a = c^2\sqrt{R}$, R = curvature scalar. This would imply a relation of the type

$$\Delta R_{\alpha\beta\gamma\delta}\,(\Delta x^{\alpha} \wedge \Delta x^{\beta})\,(\Delta x^{\gamma} \wedge \Delta x^{\delta}) \geq \hbar G/c^3. \tag{19}$$

So a general relativistic upper limit on acceleration due to quantum effects (an operational way of measuring curvature tensor) would also imply a maximal curvature and hence uncertainty relation involving Planck length.

Thus, in order to arrive at a quantum theory of gravitation, for which no limitations but only genuine uncertainty relations occur, one has to decouple metrical background and quantized (gravitational) field, i.e. to modify General Relativity Theory.

In terms of tetrads ($h^a_{\mu}\,h^b_{\nu}\,\eta_{ab} = g_{\mu\nu}$) we can define the fundamental 1-forms θ^i:

$$\theta^i = h^i_{\mu}\,dx^{\mu} \tag{20}$$

and the connection 1-forms ω^i_j:

$$\omega^i_j = \gamma^i_{jk}\,\theta^k \tag{21}$$

where γ^i_{jk} are Ricci rotation coefficients ($\gamma^i_{jk} = -\,h^i_{\mu;\nu}\,h^{\mu}_j h^{\nu}_k$) and, as it is well known, we have the relations:

$$h^a_{\alpha}x^{\beta} - x^{\beta}h^a_{\alpha} = \delta^{\beta}_{\alpha}\,(\hbar G/c^3)^{1/2} \tag{22}$$

or

$$\Delta\,h^a_{\alpha}\,\Delta x^{\beta} \geq \delta^{\beta}_{\alpha}\,(\hbar G/c^3)^{1/2} \tag{23}$$

We can take covariant exterior derivatives of eqs. (20) and (21) thereby obtaining curvature and torsion 2-forms (see also below eq.(64)):

$$\Theta^i = D\theta^i = d\theta^i + \omega^i_k \wedge \theta^k = (1/2)Q^i_{jk}\theta^j \wedge \theta^k \tag{24}$$

$$\Omega^i_j = D\omega^i_j + \omega^i_k \wedge \omega^k_j = (1/2)R^i_{jkl}\theta^l \wedge \theta^k \tag{25}$$

so that we have the conjugate relation between torsion and curvature forms

$$\left[D\,\theta^i,\ D\omega_{mj}\right] = \left[\Theta^i, \Omega_{mj}\right] = \delta^i_m\,(\hbar G/c^3)^{1/2}$$

and

$$(1/2)\left[Q^i_{kl.}\,\theta^k \wedge \theta^l,\ R_{mjkl}\,\theta^j \wedge \theta^k \wedge k^l\right] = (\hbar G/c^3)\delta^i_m \tag{26}$$

We can start from these considerations: (see for instance [7]) if we wish to connect the initial and final positions of one and the same particle we cannot avoid the uncertainty associated with torsion, i.e. for a sufficiently small area element dS, uncertainty in distance between initial and final position would be $\Delta l^\mu = Q^\mu dS$ and this would induce fluctuations in distance in the metric through $\Delta l = \sqrt{g_{\mu\nu}dx^\mu dx^\nu}$. So what is important is not the point themselves but the "fluctuations" in their position, i.e. the interval between them caused by a deformation of space itself through torsion. Note that plastic deformations are induced by torsion and are different from elastic deformations considered by Sacharov (which depend only on curvature). With quantized values, these fluctuations would also manifest as metric fluctuations. Since curvature causes relative acceleration between neighbouring test particles we have momentum uncertainty related to curvature as

$$ma^\mu ds = \Delta p^\mu = mR^\mu_{\alpha\beta\gamma}\frac{dx^\alpha}{ds}dx^\beta k^\gamma = mR\frac{dS}{ds}k^\mu \tag{27}$$

So just as position fluctuations are due to torsion, momentum fluctuation are due to curvature and we can interpret quantum effects (and then uncertainty principle) as consequences of space-time deformation i.e.

$$\Delta p^\mu \cdot \Delta x_\mu = \hbar \tag{28}$$

where

$$\Delta x^\mu = Q\,c\,ds\,k^\mu \qquad \text{and} \qquad \Delta p^\mu = m\,R\frac{dS}{ds}k^\mu \tag{29}$$

(i.e. uncertainty in initial and final positions and relative acceleration between them due to space-time deformations), Q (torsion) and curvature (R) both playing simultaneous roles as conjugate variables of the geometry (gravitational field), thus enabling us to write commutation relations between curvature and torsion (analogous to $[x, p] = i\hbar$) as

$$[Q, R] = (\hbar G/c^3)^{-3/2} \tag{30}$$

or, more explicitly, as

$$\left[Q^{\alpha}_{\mu\nu}\, dx^{\mu} \wedge dx^{\nu}, R_{\beta\mu\nu\rho}\, dx^{\mu} \wedge dx^{\rho} \wedge k^{\nu}\right] = i\hbar G/c^3 \delta^{\alpha}_{\beta} \tag{31}$$

or

$$\Delta Q\, \Delta R \geq L_{Pl}^{-3} \tag{32}$$

where L_{Pl} is the Planck length.

Thus in a sense, we have written quantum commutation relations for the observables of the background geometry, rather than for the gravitational field in a fixed background as in the usual picture; thus partially removing the inconsistency between quantizing the gravitational field and not the geometry, gravity being the geometry itself!

In fact, in the usual methods of quantizing the gravitational field one has the field on a fixed background, which gives rise to an inconsistency, because the gravitational field is just geometry so that how can we separate the geometry and the field? Here, on the contrary, we have not only quantized torsion on a fixed background but we have considered both torsion and background curvature which appear as conjugate variables.

Curvature and torsion in this context appear to be conjugate variables for which we can write commutation relations like (30).

3. Torsion and Twistors

At this point we can notice some analogies between torsion and twistors, and we would like to emphasize this point because it can constitute a further mathematical basis for the Riemann-Cartan space-time.

In this connection we would like to mention that in another work [8] it has been unambiguously shown that spinor (and then torsion which is related to spin) must form the basic building block of any relativistically covariant theory. This provides the motivation for the modern attempts to formulate the concept of space-time points in terms of twistors.

In fact, in the case of twistors [9, 10, 11] the space-time point can be written as a 2 × 2 complex matrix having SL(2, C) group structure and on this complex spin vectors SL(2, C) invariant elements act under a local Lorentz transformation. In the torsion picture we know very well that the spinor gauge fields described by tetrads transform invariantly under SL(2, C) gauge invariance (under local Lorentz rotations of tetrad). Now in the case of twistors, complex spin vectors are regarded as primary, while in the case of torsion, the spinor gauge fields are the basic entities.

This analogy can be fruitful because when we look at quantization in general we shall use complex Hilbert space. Moreover, SL(2, C) is homomorphic to O(1, 3).

The so-called twistor formalism introduced by Penrose provides a different view of quantized space-time. In the usual view one imagines a space-time where points remain intact but g_{ab} becomes quantized.

In the twistor view, the space-time 'direction of motion' of twistors (a twistor representing a massless particle in motion) remains null, the points of space-time intersection between twistors are subjected to quantization, i.e., the points become 'fuzzy' or fluctuate quantum mechanically.

We straightway notice the striking analogy between this and our notion of line defects induced by torsion. In our picture space-time points are not fundamental, but the line defects are, and intersections between defects gives rise to the space-time points whose separation is now quantized in terms of the non-closure property due to torsion as $\oint Q^{\gamma}_{\alpha\beta}\, dx^{\alpha}\, dx^{\beta} \geq (\hbar G/c^3)^{1/2}$. Fluctuations in position due to torsion and in momentum due to curvature enabled us to write quantum uncertainty relations for the conjugate variables of curvature and torsion.

In the twistor picture to fix a space-time point, P, one considers trajectory or path of massless particles in free motion through P (i.e. twistor) and it is this timelike or null vector at P that is subjected to quantization.

In the case of torsion, the spin $\hbar$ is related to a timelike vector through the torsion and the resulting non-closure is subjected to quantization. The points are derived or secondary concepts in both the cases. In both the cases the spinors are more fundamental than the world vectors.

In the case of twistors, the physical world is pictured as a sort of network of line segments, each segment is thought of as an 'object' having a well defined angular momentum $n\hbar/2$. Intersections of segments give rise to space-time points, these points are then derived objects, the twistors themselves being more basic.

In the torsion case we have a network of line defects, each associated with an intrinsic spin $n\hbar/2$.

In the twistor case complex spin vectors (on which SL(2, C) invariant elements act when a local Lorentz transformation is performed) are regarded as primary. All world vectors and tensors are then considered to be built up of spin-vectors by means of tensor products, sums etc.

In the torsion picture as we have seen, the spinor gauge fields (described by tetrads) which transform invariantly under SL(2, C) gauge invariance (under local Lorentz rotations of tetrad) are the basic entities. In fact, the metric is a derived concept, at every point (described by intersection of defects) it is the expectation value of a tetrad product, i.e.

$$\langle h_a^b h_c^d \eta_{bd} \rangle = g_{ac} \tag{33}$$

Curvature tensors and other geometric quantities are built out of these tetrads. Analogously in the twistor case at every point the metric is defined as

$$g_{ab} = \langle \varepsilon_{AB} \varepsilon_{A'B'} \rangle \tag{34}$$

in terms of complex spin vectors.

The motivation for using complex numbers to define the structure of twistor space, is, on the one hand, due to the *complex* Hilbert space of quantum modes, while on the other we have probability amplitudes superposition laws, etc. in *real* space-time. Usually these two continua are conventionally regarded as independent of each other. One of the basic features of twistor theory is that at the very outset, there are attempts to merge these two continua into one, that is, the complex quantum mechanical continuum becomes woven into space-time geometry. Thus we have an interrelation between geometry and complex numbers (entities like probability amplitudes need complex numbers!). Moreover, in twistor theory rest mass is regarded as a secondary concept. In the torsion picture also it is secondary. Here the quantized vortices associated with torsion are more basic and they lead to rest mass.

There are many ways to formally introduce twistor concepts. First of all we can observe that the most important physical attributes of a system in special relativity are its four momentum p_a and its 6-angular momentum M^{ab} $(= -M^{ba})$, defined relative to some origin 0. A twistor Z^α is an object which can be described, in a suitable coordinate frame, by four complex numbers [10]

$$(Z^0, Z^1, Z^2, Z^3) \tag{35}$$

where the complex conjugate twistor is described by

$$(\overline{Z}_0, \overline{Z}_1, \overline{Z}_2, \overline{Z}_3) = (\overline{Z^2}, \overline{Z^3}, \overline{Z^0}, \overline{Z^1}) \tag{36}$$

(the symbol $\overline{Z}^2$ denoting the complex conjugate of the complex number Z^2, etc.). For a massless particle the 4-momentum and angular momentum components, in a standard Minkowski frame, are to be related to those of Z^α by

$$\begin{aligned}
p^0 &= (1/\sqrt{2})\left(Z^3\overline{Z}_1 + Z^1\overline{Z}_0\right)\\
p^1 &= (1/\sqrt{2})\left(Z^3\overline{Z}_1 - Z^2\overline{Z}_0\right)\\
p^2 &= (1/\sqrt{2})\left(-Z^2\overline{Z}_1 - Z^3\overline{Z}_0\right)\\
p^3 &= (1/\sqrt{2})\left(Z^2\overline{Z}_1 - Z^3\overline{Z}_0\right)
\end{aligned} \tag{37}$$

and

$$\begin{aligned}
M^{01} &= -M^{10} = (1/2)\left(Z^0\overline{Z}_0 - Z^1\overline{Z}_1 - Z^2\overline{Z}_2 + Z^3\overline{Z}_3\right)\\
M^{02} &= -M^{20} = (1/2)\left(Z^0\overline{Z}_1 + Z^1\overline{Z}_0 - Z^2\overline{Z}_3 - Z^3\overline{Z}_2\right)\\
M^{03} &= -M^{30} = (1/2)\left(Z^0\overline{Z}_1 - Z^1\overline{Z}_0 - Z^2\overline{Z}_3 + Z^3\overline{Z}_2\right)\\
M^{12} &= -M^{21} = (1/2)\left(Z^0\overline{Z}_1 - Z^1\overline{Z}_0 + Z^2\overline{Z}_3 - Z^3\overline{Z}_2\right)\\
M^{32} &= -M^{23} = (1/2)\left(Z^0\overline{Z}_0 - Z^1\overline{Z}_1 + Z^2\overline{Z}_2 - Z^3\overline{Z}_3\right)\\
M^{13} &= -M^{31} = (1/2)\left(Z^0\overline{Z}_1 + Z^1\overline{Z}_0 + Z^2\overline{Z}_3 + Z^3\overline{Z}_2\right)\\
M^{00} &= M^{11} = M^{22} = M^{33} = 0
\end{aligned} \tag{38}$$

The most directly physical one, relevant for our purpose, is to use angular momentum of zero rest mass particle. The total momentum 4-vector $\overline{p}_a$ and angular momentum bivector $M^{ab} = -M^{ba}$, with respect to some origin O correspond to those of a zero rest mass particle of helicity h. We have

$$\overline{S}_a = h\,\overline{p}_a \tag{39}$$

$\overline{S}_a$ being the Pauli-Lubanski vector, being given by

$$\overline{S}_a = (1/2)\,\varepsilon_{abcd}\overline{p}^b M^{cd} \tag{40}$$

$\overline{p}^a$ is null and future pointing i.e. $pp^a = 0$ and $p^0 > 0$. Equation (39) is automatically satisfied by (37), (38) with

$$h = (1/4)\left(Z^{\alpha}\bar{Z}_{\alpha} + \bar{Z}_{\alpha}Z^{\alpha}\right) \tag{41}$$

In obtaining (41), it has not been assumed that Z^{α} and $\bar{Z}_{\alpha}$ commute but that either all components commute (classical twistor variables) or that the twistor commutation rules

$$\left[Z^{\alpha}, Z^{\beta}\right] = 0, \quad [\bar{Z}_{\alpha}, \bar{Z}_{\beta}] = 0, \quad \left[Z^{\alpha}, \bar{Z}_{\beta}\right] = \hbar\delta^{\alpha}_{\beta} \tag{42}$$

hold. (see also (66) and (72))

The two-spinor expression $p_{AA'}$ for p_a has the form

$$p_{AA'} = \bar{\pi}_A \ \pi_{A'} \tag{43}$$

(for some dual conjugate spin vector π_A).

The angular momentum in this notation is defined as [9]

$$M^{AA'BB'} = i\,\omega^{(A-B)}\pi\,\varepsilon^{A'B'} - i\varepsilon^{AB}\,\overline{\omega}^{(A'B')}\,\pi \tag{44}$$

The conjugate pair (ω^A, π_A) uniquely defines momentum p_a and angular momentum M^{ab}.

The gauge freedom in ω^A and π_A is of overall phase change, i.e. invariance under

$$\omega^{A'} \Rightarrow e^{i\theta}\,\omega^{A}, \pi_{A'} \Rightarrow e^{i\theta}\,\pi_A \tag{45}$$

ω^A, π_A are 2 spinor parts of twistors Z^{α}, written as:

$$Z^{\alpha} \Leftrightarrow \left(\omega^{A}, \pi_A\right) \tag{46}$$

The complex conjugate of twistors Z^a, is the dual $\bar{Z}_{\alpha}$, i.e.

$$\bar{Z}^{\alpha} \Leftrightarrow \left(\bar{\pi}_{A'}, \overline{\omega}^{A'}\right) \tag{47}$$

The scalar product $Z^a\bar{Z}_{\alpha}$ can be now formed as:

$$Z^{\alpha}\bar{Z}_{\alpha} = \omega^{A}\,\bar{\pi}_A + \pi_{A'}\,\overline{\omega}^{A'} \tag{48}$$

This so-called twistor norm, i.e. the invariant Hermitian form $Z^{\alpha}\bar{Z}$ has an interesting physical interpretation. It measures the helicity of a zero mass particle i.e. (from (39), (40), (41), (42)):

$$Z^{\alpha}\bar{Z}_{\alpha} = 2h \tag{49}$$

The helicity (or spin) is thus a conformal invariant. When h = 0 we have:

$$Z^{\alpha}\bar{Z}_{\alpha}=0 \tag{50}$$

and Z^{α} (or $\bar{Z}_{\alpha}$) is referred to as a null twistor.

We can think of a complexified space-time co-ordinate as

$$Z_{\mu}=x_{\mu}+ih_{\mu} \tag{51}$$

here ih_{μ} corresponds to the 'internal helicity' projection? Vector attached to the space-time point x_{μ}.

In fact, the anisotropic feature of the h-space, leads us to consider it as an attached 'projection vector' to the space-time point x_{μ} and with twistor geometry this 'projection vector' is automatically interpreted as an internal helicity; then the space-time point x_{μ} can be written as a 2 × 2 complex matrix having SL(2, C) group structure. Physically the two opposite orientations of the projection vector give rise to two opposite internal helicity ($h\,p_a$) which correspond to particle and antiparticle in isospin space and which can easily be formulated in terms of the extended metric $g_{\mu\nu}(x, \upsilon, \bar{\upsilon})$ (see below) where υ and $\bar{\upsilon}$ are two-component spinorial variables (see eq.(53) below).

Considering a complex manifold, the coordinates in this extended space where we identify the coordinate in the complex manifold as $Z^{\mu}=x^{\mu}+ih^{\mu}$, where $h^{\mu}=(1/2)\,\eta^{\mu}_{a}\,\upsilon^{a}$, can be written as

$$Z^{\mu}=x^{\mu}+(i/2)\,\eta^{\mu}_{a}\,\upsilon^{a} \tag{52}$$

where υ^{μ} (= 1, 2) is a two component spinor or, in matrix representation:

$$Z^{BB'}=x^{BB'}+(i/2)\,\eta^{BB'}_{a}\,\upsilon^{a} \tag{53}$$

where

$$x^{BB'}\ (1/\sqrt{2})\begin{bmatrix} x^{0}-x^{1} & x^{2}+ix^{3} \\ x^{2}-ix^{3} & x^{0}+x^{1} \end{bmatrix} \tag{54}$$

$$\eta^{BB'}_{a}\in SL(2, C) \tag{55}$$

The twistor equation in terms of the complex conjugate spinors $\bar{\pi}_{B}$ ($\pi_{B'}$) is

$$\bar{Z}_{\alpha}Z^{\alpha}+\eta^{BB'}_{a}\,\upsilon^{a}\,\bar{\pi}_{B}\,\pi_{B'}=0 \tag{56}$$

The four momentum corresponds to

$$\bar{p}^{BB'}=\bar{\pi}^{B}\,\pi^{B'} \tag{57}$$

$$Z_{\alpha}=(\omega^{B}, \pi_{B'});\ \bar{Z}^{\alpha}=(\bar{\pi}_{B}, \bar{\omega}^{B'})$$

with

$$\omega^{B} = i\left(x^{BB'} + (i/2)\,\eta_{a}^{BB'}\,\upsilon^{a}\right)\pi_{B'} \tag{58}$$

The helicity operator is then

$$h = Z^{\alpha}\,\overline{Z}_{\alpha} = -\,\eta_{a}^{BB'}\,\upsilon^{a}\,\overline{\pi}_{B}\,\pi_{B'} \tag{59}$$

$\overline{p}^{\mu}$ is the canonical conjugate of x^{μ} in $Z^{\mu} = x^{\mu} + (i/2)\,\eta_{a}^{\mu}\,\upsilon^{a}$

$$\overline{p}^{BB'} = \overline{\pi}^{B}\,\pi^{B'},\quad \overline{p}_{\mu}^{2} = 0 \tag{60}$$

Defining

$$\beta = i\,\upsilon^{A'}\,\pi^{A},\quad \overline{\beta} = -\,i\,\overline{\upsilon}^{A}\,\overline{\pi}_{A'} \tag{61}$$

$$h = -\,\upsilon^{A}\,\upsilon^{A'}\,\overline{\pi}_{A}\,\pi_{A'} = \beta\,\overline{\beta} \tag{62}$$

The metric is then

$$g_{\mu\nu}(x, \upsilon, \overline{\upsilon}) = g_{\mu\nu}^{BB'}(x)\,\overline{\upsilon}_{B}\,\upsilon_{B'} \tag{63}$$

One can now construct connections (both symmetric and anti symmetric) from this metric and its derivatives. Torsion and curvature corresponding to this complexified space can in turn be constructed from these connections, as we have seen in (24) and (25), i.e.

$$D\upsilon^{i} = (1/2)\,Q_{jk}^{i}\,\upsilon^{j} \wedge \upsilon^{k},\quad D\omega_{j}^{i} = (1/2)\,R_{jkl}^{i} \wedge \upsilon^{k} \tag{64}$$

One can define the radius vector of the field which can be integrated along a curve and when this is done for a loop (of radius r), one finds that with torsion, it does not return to its initial value, i.e. it gives rise to a local translation in the internal space of the radius vector field. So the change in the intersection loop of two twistors when taken along such a loop change is given by

$$\delta Z^{\alpha} = \frac{1}{r^{2}} \int \left(d\omega_{ji}^{B} + \omega_{ki}^{B} \wedge \omega_{j}^{Bk}\right) Z^{\alpha}\,dx^{i}\,dx^{j} \geq \hbar^{1/2} n^{\alpha} \tag{65}$$

The fluctuation in the norm $\langle \delta Z^{\alpha}\ \ \delta \overline{Z}_{\alpha} \rangle \geq \hbar$ then gives a twistor geometric definition of Planck's constant. Thus, this automatically leads to the quantization of the background geometry. This may provide a basis for the twistor commutation relation

$$\left[Z^{\alpha}, Z^{\beta}\right] = 0, \quad [\bar{Z}_{\alpha}, \bar{Z}_{\beta}] = 0, \quad \left[Z^{\alpha}, \bar{Z}_{\beta}\right] \geq \hbar\, \delta^{\alpha}_{\beta} \tag{66}$$

(see eq. (42) and eq. (72)).

Eqs. (64) and (65) can constitute a first step towards quantization of space-time through twistors in terms of spinors υ and complex conjugate variable $\bar{\pi}$ or $\bar{\omega}$ which are connected with torsion and curvature; in this context twistor theory seems to constitute a mathematical-geometrical basis for the commutation relation (30).

Moreover following Crampin and Pirani we have [12]

$$\bar{Z}_{\alpha}\, dZ^{\alpha} = i\pi\left(\pi_{A'}\, dx^{AA'} + x^{AA'}\, d\pi_{A'}\right) - ix^{AA'}\, \bar{\pi}_A\, d\pi_{A'} \tag{67}$$

$$= i\bar{\pi}_A\, \pi_{A'}\, dx^{AA'} = ip_a\, dx^a$$

and then we have

$$d\bar{Z}_{\alpha} \wedge dZ^{\alpha} = idp_a \wedge dx^a \tag{68}$$

where $x^{AA'}$ is the spinor equivalent of x^a. The null geodesic is well defined if and only if $\omega^A \pi_A$ is real, which is the condition that the corresponding twistor be null. The twistor $\bar{Z}_{\alpha}$ (the complex conjugate of Z^{α}) is represented by the spinor pair $(\bar{\pi}_{A'}, \omega^{A'})$; if moreover $(\lambda^A, \mu_{A'})$ represents a twistor W, the scalar product $Z_{\alpha}W^{\alpha}$ is $\bar{\pi}_A\lambda^A + \omega^{A'}\mu_{A}$. In particular, $\bar{Z}_{\alpha}dZ^{\alpha} = \bar{\pi}_A d\omega^A + \bar{\omega}^{A'} d\pi_{A'}$, so for a null twistor we have eq. (68). Then we have

$$d\bar{Z}_{\alpha} \wedge dZ^{\alpha} = idp_a \wedge dx^a \tag{69}$$

and as shown in a previous article [7] dp_a and dx^a can be associated, respectively, with fluctuations ΔR and ΔQ in the curvature and torsion, respectively, of the background geometry. Thus

$$[\Delta Q, \Delta R] = L_{Pl}^{-3} \tag{70}$$

Therefore, analogous to the well-known quantum commutation relations

$$\left[x^{\alpha}, x^{\beta}\right] = 0, \quad [p_{\alpha}, p_{\beta}] = 0 \quad \left[x^{\alpha}, p_{\beta}\right] = \hbar\, \delta^{\alpha}_{\beta} \tag{71}$$

we have the twistor commutation relations for the geometry as

$$\left[Z^{\alpha}, Z^{\beta}\right] = 0, \quad [\bar{Z}_{\alpha}, \bar{Z}_{\beta}] = 0, \quad \left[Z^{\alpha}, \bar{Z}_{\beta}\right] = \hbar\, \delta^{\alpha}_{\beta} \tag{72}$$

(see also (42) and (66)).

Thus the uncertainty relations for Z^α, Z_β take the form

$$\left[\Delta Z^\alpha, \Delta \bar{Z}_\beta\right] = (1/2)\, \hbar\, \delta^\alpha_\beta \tag{73}$$

The above discussion has been for massless particles, the twistors norm $Z^\alpha \bar{Z}_\beta$ being physically associated with the helicity of a zero mass particle.

References

1. H.E. Cartan - Compt. Rend. **174**, 437, 593 (1922); Ann. Ecole Normale **41**, 1 (1924).
2. A. Trautman - "*Theory of Gravitation*", preprint IFT/72/25, Warsaw Uni.: read at the Symposium "*On the Development of the Physicist's Conception of Nature*", Miramare, Trieste 1972.
3. H.-H.V. Borzeszkowski and H.-J. Treder - Ann. der Physik **46**, 315 (1989); see also "*On Quantum Gravity*" PRE-EL 88-05, Potsdam 1988, and H.-H.V. Borzeszkowski and H.-J. Treder, "*The Meaning of Quantum Gravity*", Dordrecht, Reidel Publ. Comp. 1988.
4. V.de Sabbata and C. Sivaram - "*Torsion and Quantum Effects*" in "*Modern Problems of Theoretical Physics*", Festschrift for Professor D. Ivanenko, ed. by P.I. Pronin and N. Obukhov, World Sci. Singapore 1991, page 143.
5. V.de Sabbata, C. Sivaram, H.-H.V. Borzeszkowski and H.-J. Treder - Ann.der Physik **48**, 197 (1991).
6. P.A.M. Dirac - "*New directions in Physics*" J. Whiley, N.Y. 1978.
7. V.de Sabbata - Il Nuovo Cimento **107A**, 363 (1994) see also: V.de Sabbata and C.Sivaram - "*Spin and Torsion in Gravitation*" World Sci. Singapore, 1994 Chp. VIII, sections 6 and 7.
8. Yu Xin - in Proceedings 5th Marcel Grossman Meeting, Australia, 1988.
9. Roger Penrose - "*Twistor theory, its aims and achievements*" in '*Quantum Gravity*' ed. by C.J. Isham, R. Penrose and D.W. Sciama, Clarendon Press, Oxford 1975.
10. Roger Penrose - "*Twistor Theory*" in '*Cosmology and Gravitation - Spin, Torsion, Rotation and Supergravity*,' ed. by Peter G. Bergmann and Venzo de Sabbata, Plenum, Press, N.Y. 1980.
11. R. Penrose and W. Rindle - "*Spinors and space-time*" Cambridge University Press, Cambridge 1986.
12. M. Crampin and F.A.E. Pirani - in "*Relativity and Gravitation*" ed. by Charles G. Kuper and Asher Peres, Gordon and Breach N.Y. 1971, page 105.

Singularity Free Cosmology in Higher Dimensions

A. Banerjee, Ajanta Das and D. Panigrahi
Relativity and Cosmology Research Centre,
Department of Physics, Jadavpur University, Calcutta - 700 032.

1. Introduction

The search for a consistent quantum theory as well as unification of gravity with other forces of nature have recently generated much interest in working with extra spatial dimensions [1,2]. Although the extra dimensions are not observable at present presumably due to their typical dimensions being of the size of the Planck length, the standard cosmology indicates that the scale factor for extra dimensions would have been comparable with the usual three dimensional space at some epoch in the early stage of the universe. The first study of the cosmological implication of the higher dimensions was made by Chodos and Detweiler [3], who pointed out that simple solutions to Einstein's equations corresponding to Kasner type space in higher dimensions would lead to the shrinking of the extra dimension to a size of planck length, whereas the usual three space would expand. The attempt to introduce the inhomogeneity in the matter energy content was made in a series of papers very recently [4,5,6] in the context of the five dimensional space-time, where the usual three space has maximal symmetry but the extra space metric component depends on both space and time coordinates. The consideration of inhomogeneity in the matter content is justified in view of different observational evidences in recent times. However, in all the said higher dimensional models as well as in a large number of Kaluza-Klein extensions of F.R.W. universe [7,8] the bigbang singularity is unavoidable. To escape from the occurrence of such a singularity the modifications of General Relativity, introducing new fields or ascribing unusual properties to the matter content have been suggested from time to time.

Senovilla's recent discovery of the singularity-free solution [9] is quite interesting in the background of what has been discussed above because it satisfies the energy conditions and has an acceptable equation of state ($\rho = 3p$). Later Patel and Dadhich [10] extended such singularity free cylindrically symmetric space-time for stiff fluid $\rho = p$ and vacuum with $\rho = p = 0$. They have further generalized these classes of solutions to include heat flux [11] and also the massless scalar field, where the equation of state $\rho = \mu p$ is satisfied with μ lying between 3 and 4.

One may wonder whether any singularity free solution exists in higher dimensional inhomogeneous cosmology, which at the same time demonstrates dimensional reduction of the extra space during evolution leading to an effective 4D space-time at the later stage. It is indeed possible to find new classes of exact solutions in five dimensions for a generalized Senovilla type metric, which describes a perfect fluid with an equation of state $\rho = \mu p$. However, solutions are obtained for $\mu = (\sqrt{3} + 1)$, $\mu = 1$ and for vacuum ($\rho = p = 0$). It should be clearly pointed out at this stage that the initial bigbang singularity can be avoided even when the energy conditions are satisfied and vorticity is absent. This is possible due to the existence of an acceleration dependent term in the generalized Raychaudhuri equation in multidimensional space-time [12]. One should note further that the solutions discussed in this article are all having the property of dimensional reduction due to the shrinkage of the extra dimensions in course of time.

2. Field equations in 5D space-time

Senovilla type 5D metric may be written as

$$dS^2 = \cos h^{2\alpha}(kt) \cos h^{2a}(mr)\,(dt^2 - dr^2)$$

$$- \frac{1}{m^2} \cos h^{2\gamma}(kt) \cos h^{2c}(mr) \sin h^2(mr)\, d\phi^2$$

$$- \cos h^{2\beta}(kt) \cos h^{2b}(mr)\, dz^2 - \cos h^{2\delta}(kt) \cos h^{2d}(mr)\, d\psi^2 \quad (2.1)$$

where $\alpha, \beta, \gamma, \delta, a, b, c, d, m$ and k are constant parameters. The metric is globally regular for the entire range of the other 4-coordinates $-\infty < t, Z < \infty$, $0 \le r < \infty$, $0 \le \phi \le 2\pi$. The above metric satisfies the regularity condition near the axis $r \to 0$. The field equations $G^M_\nu = -8\pi T^\mu_\nu$ for the metric (2.1) are given below.

$$G^0_0 = -A^{-2}\Big[k^2 \tan h^2(kt)\,(\alpha\gamma + \alpha\beta + \alpha\delta + \beta\gamma + \gamma\delta + \beta\delta)$$

$$+ m^2(a - 2b - 3c - 2d - 1) + m^2 \tanh^2(mr)\,(-b^2 - c^2 - d^2 + ab + ac$$

$$+ ad - bc - cd - bd + b + c + d\Big]$$

$$= -8\pi T^0_0 \quad (2.2)$$

$$G_1^0 = -A^2\Big[mk\tan h(kt)\tan h(mr)(a\gamma + a\beta + a\delta + c\alpha$$

$$+ b\alpha + d\alpha - c\gamma - b\beta - d\delta) + mk\tan h(kt)\cot h(mr)(\alpha - \gamma)\Big]$$

$$= -8\pi T_1^0 \tag{2.3}$$

$$G_1^1 = -A^{-2}\Big[k^2\tan h^2(kt)(\beta^2 + \gamma^2 - \delta^2 - \alpha\beta - \alpha\gamma - \alpha\delta + \beta\gamma + \beta\delta + \gamma\delta$$

$$- \beta - \gamma - \delta) + \{k^2(\beta + \gamma + \delta) - m^2(a + b + d)\} - m^2\tan h^2(mr)(ab$$

$$+ ac + ad + bc + bd + cd)\Big]$$

$$= -8\pi T_1^1 \tag{2.4}$$

$$G_2^2 = -A^{-2}\Big[k^2\tan h^2(kt)(\beta^2 + \delta^2 + \beta\delta - \alpha - \beta - \delta) + \{k^2(\alpha + \beta + \delta)$$

$$- m^2(a + b + d)\} + m^2\tan h^2(mr)(a + b + d - b^2 - d^2 - bd)\Big]$$

$$= -8\pi T_2^2 \tag{2.5}$$

$$G_3^3 = -A^{-2}\Big[k^2\tan h^2(kt)(\gamma^2 + \delta^2 + \gamma\delta - \alpha - \gamma - \delta) + \{k^2(\alpha + \gamma + \delta)$$

$$- m^2(1 + a + 3c + 2d)\} + m^2\tan h^2(mr)(a + c + d - c^2 - d^2 - cd)\Big]$$

$$= -8\pi T_3^3 \tag{2.6}$$

$$G_4^4 = -A^{-2}\Big[k^2\tan h^2(kt)\ (\beta^2 + \gamma^2 + \beta\gamma - \alpha - \beta - \gamma) + \{k^2(\alpha + \beta + \gamma)$$

$$- m^2(1 + a + 3c + 2b)\} + m^2\tan h^2(mr)\ (a + b + c - b^2 - c^2 - bc)\Big]$$

$$= -8\pi T_4^4 \tag{2.7}$$

where $g_{00} = A^2 = \cos h^{2\alpha}(kt)\cos h^{2a}(mr)$. The components of energy momentum tensor for the perfect fluid in comoving coordinates $v^\mu = (g_{00})^{-1/2}\delta_0^\mu$ may be expressed in the usual forms

$$T_0^0 = \rho,\ T_1^1 = T_2^2 = T_3^3 = T_4^4 = -p,\ T_1^0 = 0 \tag{2.8}$$

when we apply the relations (2.8) to the field equations (2.2) – (2.7) we obtain a set of relations between the constant parameters. These are

$$\alpha = \gamma$$

$$
\begin{aligned}
\alpha(a+b+d)+b(a-b)+d(a-d)&=0\\
k^2(\alpha-\beta)+m^2(b-3c-d-1)&=0\\
(\alpha-\beta)(\alpha+\beta+\delta-1)&=0\\
(b-c)(b+c+d-1)&=0\\
(\alpha-\delta)(\alpha+\beta+\delta-1)&=0\\
k^2(\alpha-\delta)-m^2(b+3c-d+1)&=0\\
(c-d)(b+c+d-1)&=0
\end{aligned}
\tag{2.9}
$$

3. Solutions of the Field Equations for $\rho = \mu p$

One set of solutions may be obtained by choosing $(\alpha+\beta+\delta)=1$ and $b+c+d \neq 1$. We have as a result $b=c=d$ from (3.9). We can further choose

$$\frac{\alpha}{1+c}=\frac{\beta}{b}=\frac{\delta}{d}=\frac{1}{(1+3b)} \tag{3.1}$$

So finally the constant parameters in this particular case can be expressed in terms of 'b' as follows

$$\alpha=\gamma=\frac{1+b}{1+3b},\quad \beta=\delta=\frac{b}{1+3b},\quad \frac{k^2}{m^2}=(1+3b)^2,\quad b=c=d,\quad a=-\frac{2b}{1+3b}$$

These values satisfy all the necessary relations (2.9). The density and pressure may be obtained from (2.2) and any of (2.4) to (2.7) respectively. The choice of an equation of state like $\rho=\mu p$ then demands $3\beta^2=1$ so that $b=-\frac{1}{2}[1 \mp \frac{1}{\sqrt{3}}]$. The value $b=-\frac{1}{2}[1-\frac{1}{\sqrt{3}}]$ is the only physically reasonable choice for $\mu>0$. In our case $\mu=(\sqrt{3}+1)$. Now since 'b' is known all the metric components can be explicitly determined. The matter density is found to be positive and is given by

$$8\pi\rho=\frac{1}{2}m^2(5-\sqrt{3})\cos h^{-\frac{2(1+b)}{1+3b}}(mr)\cos h^{-\frac{2(1+b)}{1+3b}}(kt) \tag{3.2}$$

where

$$\frac{1+b}{1+3b}=\left(1+\frac{2}{\sqrt{3}}\right).$$

The energy density at any location is vanishingly small at $t \to -\infty$ or $t \to +\infty$ having the maximum value at $t = 0$. There is no singularity at any stage of evolution. In the solutions given above it is noted that $\beta = \delta < 0$ and since $|g_{44}| = \cosh^{2\delta}(kt)\cosh^{2d}(mr)$ the extra dimension shrinks with time.

Another set of choices of the parameters is determined by $\alpha + \beta + \delta = 1$ and also $b + c + d = 1$, so that the relation similar to (3.1) will now be

$$\frac{\alpha}{1+c} = \frac{\beta}{b} = \frac{\delta}{d} = \frac{1}{2}.$$

The remaining relations are $(a + b + c) - b^2 - c^2 - bc = 0$ and $4m^2 = k^2$. The above set of choices immediately result in an equation of state for a stiff fluid that is $\rho = p$.

$$8\pi\rho = 8\pi p = m^2(a - c - 3)\cosh^{-2(1+\alpha)}(2mt)\cosh^{-2a}(mr) \qquad (3.3)$$

In this class of solutions for a stiff fluid if any two of the constant parameters are chosen, the others may be determined. One can choose the parameters in an appropriate way so that the fifth dimension shrinks in course of time.

The matter free limit of the above set of solutions may be obtained by putting $a = (c + 3)$. Such five dimensional vacuum solutions are not unique because there is still the freedom to choose another constant parameter. One example of the vacuum metric showing contraction of the extra space along with the expansion of the usual 3-space may be given by

$$dS^2 = \cosh^2(2mt)\left\{\cosh^8(mr)\,(dt^2 - dr^2) - \frac{1}{m^2}\cosh^2(mr)\sinh^2(mr)d\phi^2\right\}$$

$$- \cosh^2(2mt)\cosh^4(mr)dz^2 - \cosh^{-2}(2mt)\cosh^{-4}(mr)d\psi^2 \qquad (3.4)$$

It is shown above that we can have five dimensional singularity-free space-time in cylindrically symmetry showing dimensional reduction for the extra space and satisfying the equation of state $\rho = \mu p$. So far we have given explicit forms of metric in three different cases $\mu = (\sqrt{3} + 1)$, $\mu = 1$ (Stiff fluid) and $\rho = p = 0$ (Vacuum). In the corresponding 4D space-time Dadhich and his collaborators have shown that the fluid solutions separate out into two classes having $\rho = 3p$ and $\rho = p$. However in 5D space-time we have greater freedom and presumably there exist more solutions both for fluid and empty space, because there are larger number of parameters to be chosen.

References

1. Appelquist, T., Chodos, A. and Freund, P.G.O. (ed.) 1987, *Modern Kaluza-Klein Theories* (Reading, MA: Addison-Wesley).
2. Witten, E. (1984) Phys. Lett., **1448**, 351.
3. Chodos, A. and Detweiler. S, (1980) Phys. Rev. **D21**, 2167.
4. Chatterjee, S. and Banerjee, A. (1993), Class. Quant. Grav. **10**, L1.
5. Chatterjee, S., Panigrahi, D. and Banerjee, A. (1994), Class. Quant. Grav. **11**, 371.
6. Banerjee, A., Panigrahi, D. and Chatterjee, S. (1994), Class. Quant. Grav. **11**, 1405.
7. Sahdev, D. (1984), Phys. Lett., **137B**, 155.
8. Ishihara, H. (1984), Prog. Theor. Phys., **72**, 376.
9. Senovilla, J.M.M., (1990), Phys. Rev. Lett., **64**, 2219.
10. Patel, L.K. and Dadhich, Naresh, 1993, March preprint - IUCAA 10/93.
11. Tikekar, Ramesh, Patel, L.K. and Dadhich, N. (1994), Gen. Rel. Grav., **26**, 647.
12. Banerjee, A., Panigrahi, D. and Chatterjee, S. (1995) J. Math. Phys. (In press).

Baryogenesis at the Electroweak Scale†

U.A. Yajnik
Physics Department, Indian Institute of Technology,
Bombay 400 076

Abstract

The realization that the electroweak anomaly can induce significant baryon number violation at high temperature and that the standard models of particle physics and cosmology contain all the ingredients needed for baryogenesis has led to vigorous search for viable models. The conclusions so far are that the Standard Model of particle physics cannot produce baryon asymmetry of required magnitude. It has too little *CP* violation and sphaleronic transitions wipe out any asymmetry produced if the Higgs is heavier than about 50 GeV, a range already excluded by accelerator experiments. We review the sphaleron solution, its connection to the high temperature anomalous rate and then summarise possibilities where phenomenologically testable extensions of the Standard Model may yet explain the baryon asymmetry of the Universe.

1. Introduction

An observed fact of nature is the asymmetry between the occurrence of matter and antimatter. This asymmetry is not of a local nature as evidenced by an almost continuous distribution of luminous bodies or hydrogen clouds in the Galaxy as also on the extragalactic scales. Violent annihilation processes that may be expected at the boundary of regions containing matter and antimatter are also not to be seen [1]. Since baryon number is a good symmetry of all observed processes, one has to assume the asymmetry to be of primordial nature and thus the problem passes into the domain of the early Universe.

The net baryon number is not an important conserved number from the point of view of elementary particle physics. It is not known to couple to any gauge bosons, which would have justified its conservation. Thus we may safely assume that some high energy processes yet undiscovered in fact violate baryon number [2]. Then in the fleeting moments of the early Universe, at ultra high

---

† Presented at Symposium on Early Universe, IIT, Madras, Dec. 1994. Work supported in part by the Department of Science and Technology.

temperatures, the number was not conserved and what we see is the residue left over after the annihilation of baryons and antibaryons. The above line of thinking suffers from a further, deeper problem when we compare the density of the net baryon number n_B with the entropy density n_γ of photons. Standard Big Bang cosmology tells us that above a certain temperature in the early Universe, baryon antibaryon pairs would be freely created by photons, and the approximate thermodynamic equilibrium would make the separate densities of the baryon number n_b and the antibaryon number $\bar{n}_b$ to be of the same order of magnitude as n_γ aside from the asymmetry induced by processes at an even higher energy scale. If the high energy processes violated baryon number freely, we would expect the asymmetry $n_B = n_b - \bar{n}_b$ to be of the same order of magnitude as n_b or $\bar{n}_b$. In that case, at lower energies, after the mutual annihilation, we expect $n_b = n_B \simeq n_\gamma$. Since both n_B and n_γ scale as S^{-3}, where S is the Friedmann scale factor, the ratio of the two should remain constant throughout the later history of the Universe. This wishful thinking is contradicted by the observed value of this ratio[1], which is in the range $10^{-10} - 10^{-12}$. This is obtained by direct estimate of the density of luminous matter and hydrogen clouds, compared to entropy density of the microwave background. There is a second method which confirms this value, as well as confirming the basic premise of the Big Bang cosmology. This is the n_B/n_γ ratio needed so that we have the nucleosynthesis data about He to H by weight at its correct value[1] of 25%. We are thus faced with the challenge of introducing particle physics interactions that violate baryon number, at the same time providing for the asymmetry, a fine tuned number in the range mentioned above.

In the following, in section 2, we recapitulate the subtle requirements for dynamical production of baryon number in the early Universe. In section 3, we introduce the sphaleron and the reasons for looking for baryogenesis at the electroweak scale. We also discuss here the bound placed on the Higgs boson mass by the requirement of electroweak baryogenesis. In section 4 we introduce a class of mechanisms fulfilling the requirements of electroweak baryogenesis. These will rely on the details of the electroweak phase transition, to be understood in terms of the temperature dependent effective potential. In section 5 we discuss the work done by our group related to previous section, viz. baryogenesis in electroweak phase transition induced by cosmic strings. Section 6 contains the conclusions. Due to limitations of space this review is rather brief and selective. The references cited contain more details. I hope however to convey the essentials and to share the excitement associated with making just enough baryons to ensure that we exist in the state we do.

2. General Requirements

The possibility of cosmological explanation of baryon asymmetry relying on particle physics was proposed by Sakharov[3] in the early days of *CP* violation as well as Microwave Background. Suppose there exist reactions in which baryon number (B) is violated. However if there is charge conjugation symmetry (C), such reactions cannot give rise to net baryon number, since both particles as well as antiparticles would be equally created or destroyed. Let us also suppose that *C* is violated in such reactions. However, as the reaction products begin to build up, the reverse reactions would also become viable, returning the products to the *B* symmetric state. We must therefore also suppose existence of different rates for the forward and the reverse reactions, a possibility realisable in particle physics theories with violation of time reversal symmetry *T*, or equivalently, *CP* assuming CPT invariance, *P* being parity. However, in thermal equilibrium, *CPT invariance still implies equality of* n_b *and* $\bar{n}_b$. Therefore one also needs out-of-equilibrium conditions. In the early Universe, these could be provided by the decay or decoupling of particles as certain temperature thresholds are crossed, or by the occurrence of a phase transition due to formation of condensates. Then with the state of the Universe not time symmetric, the time irreversible processes could leave their distinct mark.

It is clear that several factors have to conspire rather delicately to produce the required results. Our hope is that we make hypotheses that are generic and yet lead to this rather fine tuned number n_B/n_γ. Ideally we would like to allow for maximal possible *B* violation as well as maximal possible *CP* violation and have the small number come out compulsively as the result of a distinct, preferably unique mechanism. Proposals of this nature were made in the context of grand unified theories, where out-of-equilibrium decays of superheavy bosons led naturally to the number needed. At present we have no compelling model of grand unified interactions, but that is not the reason why we shall turn to electroweak baryogenesis. It is in fact deeper understanding of rather intricate facts of the Standard Model itself that lead us to look at this energy scale in greater detail.

3. The anomaly and the sphaleron

In Quantum Mechanics we usually expect the symmetries of the classical system to be reflected as linear invariances of the Hibert space, and a conserved quantity is expected to be represented by a hermitian operator commuting with the Hamiltonian. This however is not always true and a variety of other possibilities is now known to occur for the case of Relativistic Field Theory. In the phenomenon known as anomaly [4], a classically conserved axial vector current associated with fermions may turn out to be not conserved in the Quantum Theory. Specifically, one finds that

$$\partial_\mu j_A^\mu = \frac{g^2}{32\pi^2}\, \varepsilon^{\mu\nu\rho\sigma} F_{\mu\nu} F_{\rho\sigma} \tag{1}$$

where g is the gauge coupling. The anomaly of the fermionic current is associated with another interesting fact of gauge field theory. It was shown by Jackiw and Rebbi that the ground state of a nonabelian gauge theory consists of any configurations of gauge fields which although not permitting any nonzero physical field strengths, can not be continuously gauge transformed into each other, *i.e.*, the gauge transformation connecting them cannot be deformed to the identity transformation. Such pure gauge vacuum configurations can be distinguished from each other by a topological charge called the Chern-Simons number.

$$N_{C-S} = -\frac{2}{3}\frac{g^3}{32\pi^2} \int d^3x\, \varepsilon^{ijk} \varepsilon^{abc} A_\nu^a A_\rho^b A_\sigma^c \tag{2}$$

The existence of gauge equivalent sectors labelled by the Chern-Simons number is related to the fermionic anomaly because one can show that the RHS of eqn.(1) is equal to a total divergence $\partial_\mu K^\mu$ where

$$K^\mu = \varepsilon^{\mu\nu\rho\sigma} \left(F_{\nu\rho}^a A_\sigma^a - \frac{2}{3} g\, \varepsilon^{abc} A_\nu^a A_\rho^b A_\sigma^c\right) \tag{3}$$

so that with $F^{\mu\nu} = 0$,

$$\Delta Q_A \equiv \Delta \int j_A^0\, d^3x = -\Delta \left(\frac{g^2}{32\pi^2} \int K^0\, d^3x \right) \equiv -\Delta N_{C-S} \tag{4}$$

Thus the violation of the axial charge by unit occurs because of a quantum transition from one pure gauge configuration to another. The standard model sphaleron[5] is supposed to be a time independent configuration of gauge and Higgs fields which has maximum energy along a minimal path joining sectors differing by unit Chern-Simons number. See figure 1.

It is convenient to obtain this time independent solution in the approximation that the Weinberg angle is zero. Thus, the sphaleron of an $SU(2)$ theory spontaneously broken by a complex isospinor Higgs is given by[5][6] (in the gauge $A_0^a = 0$)

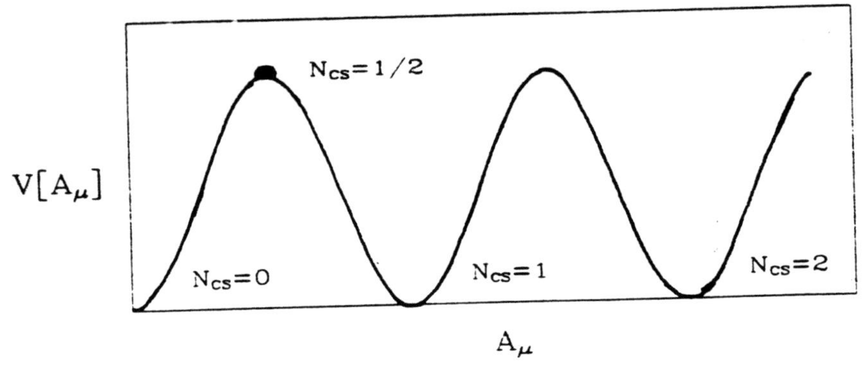

Fig. 1 Energy profile of gauge fields

$$\sigma^a A_i^a = -\frac{2i}{g} f(r) \frac{\partial}{\partial x^i} U^\infty(\vec{r}) (U^\infty(\vec{r}))^{-1} \tag{5}$$

and

$$\phi(\vec{r}) = h(r)\, U^\infty(\vec{r}) \begin{pmatrix} 0 \\ \mu \end{pmatrix} \tag{6}$$

with

$$U^\infty(\vec{r}) = \frac{1}{r} \begin{pmatrix} z & x+iy \\ -x+iy & z \end{pmatrix} \tag{7}$$

The energy of such configurations can be estimated to be $M_w/\alpha_w \simeq 10$ TeV. Detailed study[7][24] of the sphaleron with correct value of the Weinberg angle does not change the conclusions to be elaborated below. The small mixing with the $U(1)_Y$ gives rise to a small magnetic dipole moment to the sphaleron, with accompanying modification in the energy.

In the Standard Model, Q_A turns out to be the combination of the baryon and lepton numbers, $B+L$. If we assume the physical vacuum to be a state characterised by a definite value of N_{C-S}, for instance the valley at $N_{C-S}=0$ in fig.1, a spontaneous quantum transition to another state of $\Delta N_{C-S} = \pm 1$ can occur only by passing under an energy barrier of height at least as much as set by the sphaleron energy $E_{sph} \simeq 10$ TeV.

Kuzmin, Rubakov and Shaposhnikov[8] conjectured that at high temperatures, the sphaleron occurs freely as a fluctuation, and that the system makes transitions to neighboring valleys by going *over* the barrier. This would establish a chemical equilibrium between baryons and antibaryons (as well as leptons and antileptons). Any preexisting asymmetry in the $B+L$ number would therefore be wiped out at the Weinberg-Salam phase transition scale.

Subsequent analysis has substantiated this conjecture in two different temperature ranges: i) $0 << T << E_{sph}$ and ii) $T \geq E_{sph}$ using different techniques. Case i) is amenable to reliable approximation techniques.[9] Accordingly, the thermal rate for unit change in Chern-Simons number is

$$\frac{\Gamma}{V} = \frac{T^4 \omega}{M_w(T)} \left(\frac{\alpha w}{4\pi}\right)^4 N_{tr} N_{rot} \left(\frac{2M_w(T)}{\alpha_w T}\right)^7 \exp\left\{\frac{-E_{sph}(T)}{kT}\right\} k \tag{8}$$

Here $\omega = \partial^2 V_{eff}/\partial\phi^2 \,|_{\phi=0}$; N_{tr} and N_{rot} are counts of sphaleron zero modes estimated to be $N_{tr} \times N_{rot} \cong 1.3 \times 10^5$ and $k \tilde{} 1$ is a determinant. For case ii), no approximation techniques exist. Sphaleron does not exist because $\langle \phi \rangle^T = 0$ in the high temperature regime. But heuristic arguments suggest[10]

$$\Gamma = A(\alpha_w T)^4 \tag{9}$$

where A is a dimensionless constant which should be close to unity. We shall refer to this as the *high temperature* mechanism. In order to establish this mechanism simulations have been carried out on a lattice[11]. Gauge fields are set up on a lattice in contact with a heat bath and allowed to evolve in fixed time steps, keeping track of the integral number Q_{C-S} at every stage. The one such calculation carried out[11] indeed detects occasional rapid jumps signalling $\Delta Q_{C-S} = \pm 1$, and an empirical value of A between 0.1 and 1.0. See fig.2.

Γ is the anomalous transition rate ignoring the presence of fermions, and is equal for $\Delta Q_{C-S} = +1$ and $\Delta Q_{C-S} = -1$. The rate of B-number violation is then obtained from the difference between the forward and reverse rates, which depend upon the chemical potentials of the baryons and the antibaryons. Then the final result is

$$\frac{\partial(B+L)}{\partial t} = -\frac{13}{2} n_f \frac{\Gamma}{T^3} (B+L) \tag{10}$$

where n_f is the number of fermion generations. The general conclusion is therefore, that the anomaly is unsuppressed at high temperatures and no net $B+L$ can remain. This gives rise to two broad options for explaining the baryon asymmetry: A) The $B-L$ number of the Universe is non-zero for some reason, so that with

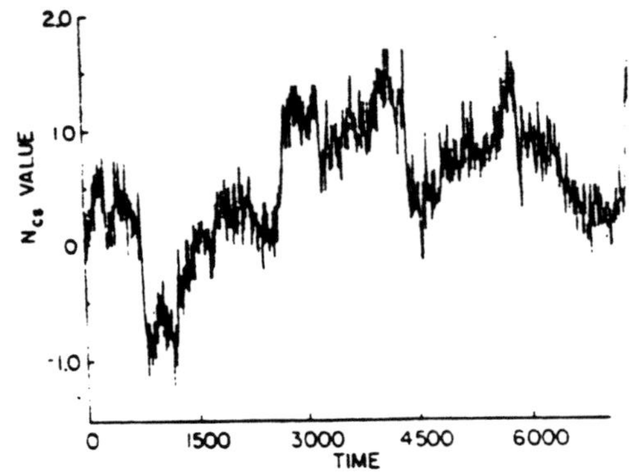

Fig.2 N_{C-S} evolution in a thermal bath

$B+L=0$, $n_B = n_L \neq 0$ survives. For this to be a satisfactory explanation, one needs a natural mechanism for non-zero n_{B-L}. B) Net $B+L$ is generated at a scale not much larger than the electroweak scale, and is neutralized incompletely by high temperature electroweak processes. We shall not take up a discussion of these possibilities but pursue the possibility of baryogenesis *at* the electroweak scale. Hopefully, the latter approach has fewer free parameters and will be easy to check against phenomenology.

There is at least one important consequence of the above analysis which we can derive within known phenomenology. Suppose the net B-number just after the electroweak phase transition is B_{in}. Since the rate in eq.(8) is known, we may integrate eq.(10) to calculate how much B-number survives the menace of the sphaleron. We note that the rate Γ depends on the temperature dependent expectation value $\langle \phi \rangle^T \equiv v$, which in turn is determined by the parameters of the Higgs potential, and hence in turn by the Higgs mass m_H. If the rate Γ is slow enough to become comparable to the expansion rate of the Universe, the sphalerons will not succeed in neutralising the $B+L$ number. Following Shaposhnikov[12], we integrate eq.(10) and obtain for suppression factor $S \equiv B_0/B_{in}$,

$$S = \exp\left(-\Gamma/H\right) \tag{11}$$

where B_0 is the net baryon number left over, and H is the Hubble parameter just after the electroweak scale. S has an implicit dependence on m_H, which is plotted in fig.3.

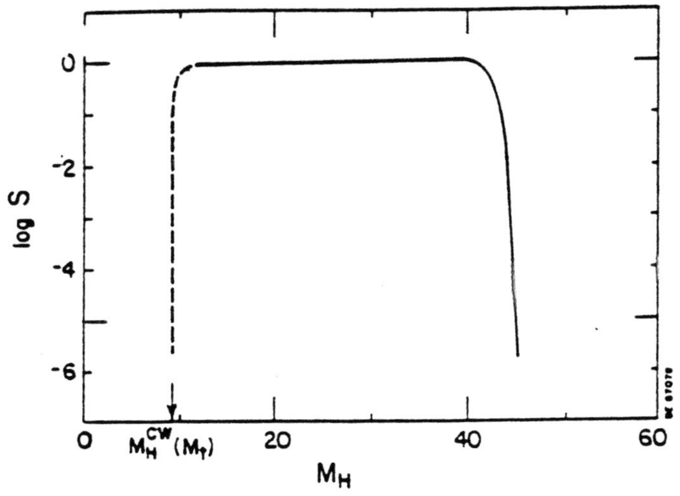

Fig.3 Suppression factor vs. Higgs mass

We see that for large m_H such as $m_H > 70$GeV, $B+L$ would have to be zero *regardless of the physics beyond the electroweak scale*. A light Higgs mass $m_H \sim 20$ to 35GeV allows all the asymmetry before the electroweak phase transition to survive. Assuming a modest value $B_{in} = 10^{-5}$, we are led to the conclusion $m_H < 45$GeV since B_0 must be 10^{-10}. Since accelerator experiments have already ruled out m_H lighter than 57GeV, this raises serious doubts about the completeness of the Standard Model with one Higgs. This is indeed the most significant result implied by the anomaly structure of the Standard Model.

4. Models of Electroweak Baryogenesis

It is clear from the preceding section that the Standard Model needs to be modified, firstly to generate baryon asymmetry and secondly, to prevent the wash out of the asymmetry. We shall consider here some modifications to the Standard Model that are minimal and satisfy the above requirements. In particular, we shall examine models in which the asymmetry is generated *at* the electroweak phase transition.

As was explained in section 2, time irreversible processes are an essential ingredient of any recipe for baryogenesis. Kuzmin et al[8] pointed out that this requirement could be met neutrally if the electroweak phase transition was first order. Here by first order we mean one in which the order parameter changes discontinuously at the phase transition. The free energy of the Higgs field is given in the field theoretic formalism by the finite temperature effective potential. The high temperature expansion for the same is given correct to $O(\hbar)$ by

$$V_{eff}^T = -(2\lambda\sigma^2 + M_1^2 - (M_2/\sigma)^2 T^2)\,\phi^2 - \frac{T}{4\pi}\left(\frac{M_3}{\sigma}\right)^3 \phi^3 + \lambda\phi^4 + \left(\frac{M}{\sigma}\right)^2 \phi^4 \ln\left(\frac{\phi}{\sigma}\right)^2 \quad (12)$$

where M_1, M_2 and M_3 are mass dimension parameters depending on physical masses M_w, M_z, M_t; σ is the zero temperature expectation value of the Higgs field, $\sigma = 246\text{GeV}$; λ determines the Higgs self-coupling. The opposite signs between T^2 and the zero-temperature coefficient in the ϕ^2 term signals that for large enough temperature, the effective mass-squared of the Higgs is positive and the symmetry no longer appears broken[13]. The form of the quartic potential leads to a variation in its shape with change in the parameter T as shown in fig.4.

There exists a temperature T_1 at which the system has two equienergetic minima separated by a barrier. This barrier persists till $T_c < T_1$, at which the second derivative of V at $\phi = 0$ changes sign from positive to negative so that $\phi = 0$ no longer remains a minimum. Between the temperatures T_1 and T_c, the system is normally at $\phi = 0$ since that is the condition persisting from $T > T_1$. However, since

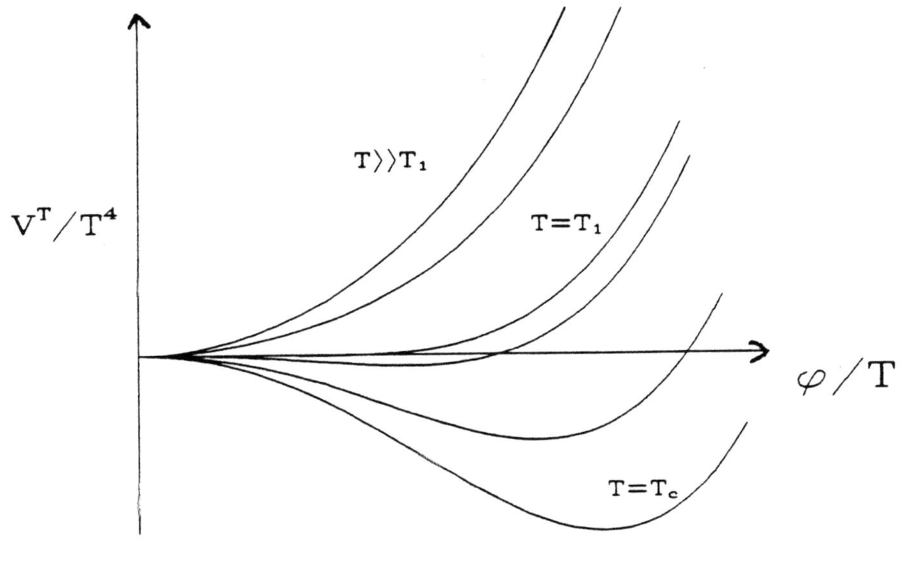

Fig.4 Variation in V_{eff}^T with T

$\phi_2 \equiv \phi \neq 0$ is favorable, thermal fluctuations and quantum tunnelling across the barrier is possible. Whenever tunnelling to the true vacuum occurs in any region of space, it results in a "bubble", the inside of which consists of the true vacuum ϕ_2 and outside is still the unconverted false vacuum. According to a well developed formalism[14][15], the tunnelling probability per unit volume per unit time is given by

$$\gamma = C T^4 e^{-S_{bubble}} \tag{13}$$

where S_{bubble} is value of

$$S = 4\pi \int r^2 dr \left\{ \frac{1}{2} \phi'^{\,2} + V_{eff}^T[\phi] \right\} \tag{14}$$

extremised over ϕ configurations which satisfy the "bubble" boundary conditions $\phi(r = 0) = \phi_2$, $\phi \to 0$ as $r \to \infty$. Once a bubble forms, energetics dictates that it keeps expanding, converting more of the medium to the true vacuum. The expansion is irreversible and provides one of the requisite conditions for producing baryon asymmetry. The important question is whether the phase transition in the electroweak theory is first order or second order. Details calculations support the view that the form of the potential is indeed as given above giving rise to a temperature $T_1 > T_c$, so that the phase transition is first order. We thus have a B-number violating mechanism, an irreversible process as well as the well known *CP* violating effects right within the Standard Model, thus giving rise to the hopes of explaining the Baryon asymmetry at the electroweak scale.

Before proceeding with this discussion we should note that first order phase transition with bubble formation is not the only way time asymmetric conditions can arise. It has been pointed out by Brandenberger and collaborators that even with a second order phase transition, cosmic strings can play an important role in catalysing baryon asymmetry production if other favorable conditions exist.[16] We can not include this interesting possibility for want of space.

The hope expressed above of explaining B asymmetry within the Standard Model is quickly belied by the fact that the extent of known *CP* violation is too small. A model independent dimensionless parameter characterising the scale of this effect has the value[17] $\delta_{CP} \sim 10^{-20}$. Since such a factor is expected to appear multiplicatively in the final answer, the resulting asymmetry would be too small. Additionally, we saw that the Standard Model Higgs seems to erase any B asymmetry generated prior to the electroweak scale. This leads us to make the minimal extension to the Standard Model, viz., to include one more complex Higgs doublet. The possibility of such has been extensively considered in other contexts as well[18]. For our purpose, this is a good extension to consider for two reasons 1) a phase transition with two Higgs doublets has the possibility of not wiping out the produced baryon asymmetry and still allowing the lightest Higgs to be heavier than 60 GeV[19]. 2) it is a source of additional *CP* violation which does not conflict with any known phenomenon.[18]

In the following we shall review one of the proposed scenarios for electroweak baryogenesis in some detail, and refer to reader to detailed reviews[20] for other possibilities. One class of possibilities we are unable to take up is that due to Cohen, Kaplan and Nelson[21].

There are several proposals along these lines[20]. Unfortunately we cannot include any details of most. One class of proposals by Cohen, Kaplan and Nelson involves scattering of neutrinos from the walls of expanding bubbles. If the neutrino is massive and has majorana mass, lepton number violation can occur in such a scattering, biasing the L number density in front of the wall, after *CP* violation has been taken into accopunt. Outside the wall, the high temperature anomalous process would be going full swing, setting the $B+L$ number to zero, thereby creating $B=-L$, i.e., negative of the L generated by wall reflections. On this basic scheme several phenomenologically viable models have been proposed[21].

In the present review we shall treat in some detail only one class of models proposed by McLerran, Shaposhnikov, Turok and Voloshin[22]. Consider the model with two Higgs doublets. In the course of the phase transition, both of these acquire nonzero vacuum expectation value. Being complex, their expectation values would generically differ in their phase, thus allowing *CP* violation in their nontrivial ground state. The bias towards creation of baryons as against antibaryons would be signalled by the presence of terms in the effective action which are linear in the Chern-Simmons number. Net baryon production can result only if *CP* violating effects are coupled to this biasing term. A term with appropriate properties is contributed by the triangle diagram shown in fig.5.

The presence of two Higgs rises the danger of flavour changing neutral currents, which is usually circumvented by coupling only one of the Higgs to the fermions or coupling up type fermions to one and down type to the other[23]. In either case, we get the dominant contribution to above kind of diagram only from a top quark loop with both scalar external legs coupled to the same Higgs. The $T\neq 0$ correction from this diagram can be calculated to be

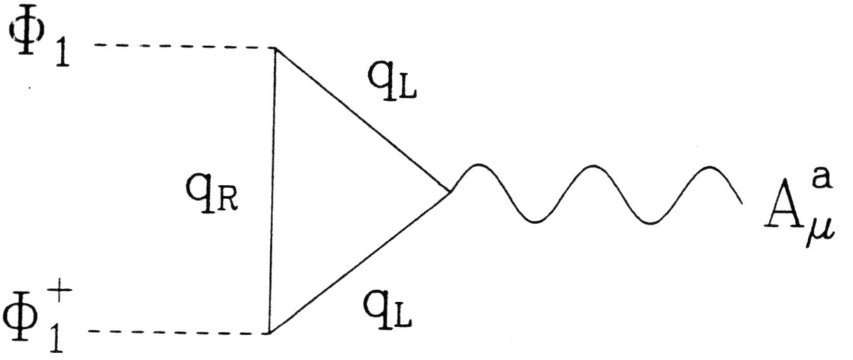

Fig.5 A nontrivial contribution to the S_{eff}

$$\Delta S = \frac{-7}{4}\zeta(3)\left(\frac{m_t}{\pi T}\right)^2 \frac{g}{16\pi^2}\frac{1}{v_1^2}$$

$$\times \int (\mathcal{D}_i\,\phi_1^\dagger\,\sigma^a\,\mathcal{D}_0\,\phi_1 - \mathcal{D}_0\,\phi_1^\dagger\,\sigma^a\,\mathcal{D}_i\,\phi_1)\,\varepsilon^{ijk}F^a_{jk}\,d^4x \qquad (15)$$

where m_t is the top quark mass, ζ is the Riemann zeta function, and the σ^a are the Pauli matrices. For homogeneous but time varying configurations of the Higgs fields, in the gauge $A_0^a = 0$, we can rewrite this in the form

$$\Delta S = \frac{-i7}{4}\zeta(3)\left(\frac{m_t}{\pi T}\right)^2 \frac{2}{v_1^2}$$

$$\int dt\,[\phi_1^\dagger\,\mathcal{D}_0\,\phi_1 - (\mathcal{D}_0\,\phi_1)^\dagger\,\phi_1\,]\,N_{CS} \qquad (16)$$

$$\equiv \mathcal{O}N_{CS}$$

which has the linear dependence on the N_{CS} as required. The expectation value of the operator $\mathcal{O}$, the imaginary part of $\phi_1^\dagger\,\mathcal{D}_0\,\phi_1$, acts as the chemical potential for this number. $\langle\,\mathcal{O}\,\rangle$ is nonzero only in the walls of bubbles, which is what we need. However, to lowest adiabatic order $\mathrm{Im}\phi_1$ can be made zero by choice of gauge, and this persists when first derivatives are taken. In the next adiabatic order, one finds

$$\mathcal{D}_\mu \mathcal{D}^\mu \, \mathrm{Im}\phi_1 = \frac{1}{2} R^3 \cos\gamma \, (\lambda_5 \cos\gamma \sin\gamma \sin\xi - \lambda_6 \sin^2\gamma \sin 2\xi) \qquad (17)$$

where $\langle \phi_1 \rangle \sim \langle \phi_2 \rangle \sim R$ in the translation invariant ground state, γ and ξ are angles specifying relative phases of $\langle \phi_1 \rangle$ and $\langle \phi_2 \rangle$; λ_5, λ_6 are dimensionless quartic couplings in the two-Higgs doublet theory[18]. ξ characterises the *CP* violation which will show up only in the scalar sector. To arrive at the numerical estimate, we take $\mathcal{D}_\mu \mathcal{D}^\mu \sim M_H^2(T)$, the temperature dependent Higgs mass-squared, which also sets the scale for the bubble wall thickness. This leads to[22] $n_B/n_\gamma \sim 10^{-3} \alpha_w^4 \sin 2\xi (T_c)$. If the quartic couplings as well as $\sin 2\xi (T_c)$ are all $\mathcal{O}(1)$, this leads to an answer in the correct range of values.

The bubble profile can be computed by making reasonable ansatz and the above calculation can be done numerically. In a particular case of bubble formation[24][28], one finds wall thickness $\sim 40T^{-1}$ and n_B/n_γ indeed $\sim 10^{-9}$.

5. String induced phase transition

For all mechanisms relying on the first order nature of the phase transition, the thickness and speed of the bubble walls are crucial parameters. Some of the mechanisms would work only in thin fast walls and others only for thick walls[20][25]. It is possible that the electroweak phase transition was induced by cosmic strings present from an earlier symmetry breaking transition. That this is possible for a generic unified theory with several stages of symmetry breakdown was shown in ref.[26]. This was investigated in detail for the electroweak effective potential in [27], where it is shown that the thickness of bubble walls in this case is

$$\Delta r = s(m_H)\, T^{-1} \qquad (18)$$

where $s(m_H)$ is a scaling factor which varies in the range $0.7 - 0.5 \times (m_H/GeV)$ as m_H varies from 60 to 120 GeV. For the wall velocity we find $v \sim 0.5$ for the same range of Higgs mass. A multiple time snapshot of the progressing bubble wall is shown in fig.6.

This mechanism invokes the existence of new gauge forces at higher energies. However, the wall parameters given above are determined entirely by the standard model physics, viz., m_H. These results show that the walls of string induced bubbled provide adiabatic conditions for a B-asymmetry generating mechanism, in particular conditions quite suited for the operation of the McLerran-Shaposhnikov-Turok-Voloshin (MSTV)[22] mechanism.

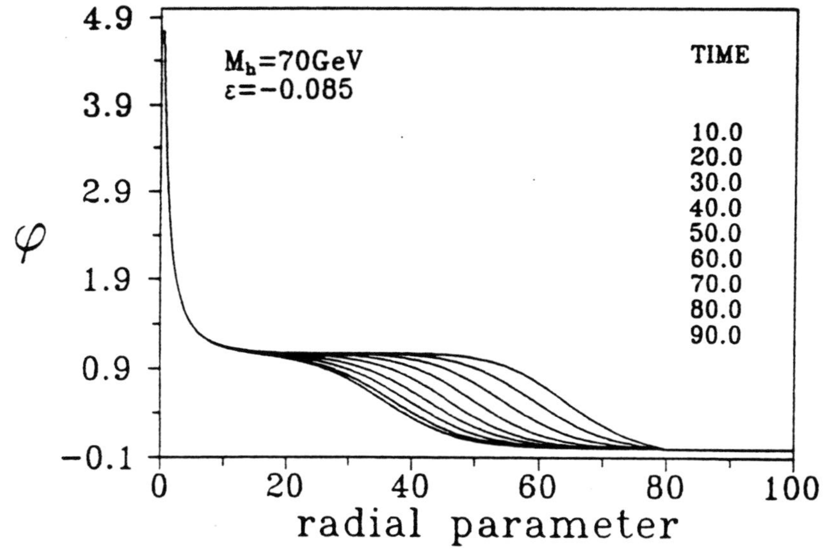

Fig.6 The induced bubble solution

MSTV mechanism suffers from the drawback that to first adiabatic order in the spatial variation of the Higgs fields, it does not produce B-asymmetry. This happens because *CP* violation comes into play only if both the Higgs are involved whereas the FCNC constraint forces coupling of the top quark only to one of the two. Recently it has been shown[29] that the Glashow-Weinberg criterion is sufficient but too strong, and that the possibility of a fermion coupling to a small extent to another Higgs is open. Accordingly we investigated[30] the MSTV mechanism with this extension in the Yukawa couplings. In this case two additional diagrams similar to the one in fig.5 contribute to S_{eff}. The value of the operator $\mathcal{O}$ was commuted in string induced bubble walls. The results are numerically in the same range; this is to be expected since the FCNC constraint still keeps the contribution of additional diagrams small but the effect is in the first adiabatic order, hence more robust.

6. Conclusion

For baryogenesis to occur in the early universe, three conditions of Sakharov are necessary. In the Standard Model, anomalous nature of the $B + L$ current allows for the violation of this number. Further, the understanding of the sphaleron solution permits the calculation of the rate of violation of this number at high temperature, indicating that the rate of violation becomes significant near the phase transition scale. Numerical simulations also suggest that the violation is completely unsuppressed above electroweak symmetry breaking scale. Secondly, *CP* violating interactions are possible in simple extensions of the Standard Model, although the *CP* violation is too small to produce the observed B-asymmetry. Finally, upon investigating the electroweak effective potential at high temperature, it is found to suggest a first order phase transition, thus providing the out-of-equilibrium conditions required by Sakharov's criteria. This raises the possibility that all the observed baryon excess in the Universe was manufactured at the electroweak scales and mostly involving known physics. We have reviewed the MSTV[22] mechanism involving two $SU(2)$ doublet scalars working as Higgs particles. We find several variations of this mechanism that would also be effective, in particular the one with less stringent restriction on Yukawa couplings enhances this effect.

The highly effective baryon-number violation suggested by the sphaleron and by numerical simulations above the electroweak phase transition raise the spectre of a universe without baryons if some mechanism for guarding them against the sphaleron menace does not exist. Shaposnikov's work duly extended shows that this implies that the Higgs mass in the Weinberg-Salam theory must be less than about 50 GeV, a range already excluded by accelerator experiments. This strongly suggests that the Standard Model needs an extension. This is very valuable information one derives about the fundamental forces at the microscopic scale by studying the early Universe.

Acknowledgement

I would like thank the organisers for the hospitality. The research and travel were made possible by a DST sponsored project.

References

1. Kolb E.W. and Turner M.S., "*The Early Universe*", Addison-Wesley (1990).
2. Weinberg S., in "*Lectures on Particles and Fields*", edited by K. Johnson et al, (Prentice-Hall, Englewood Cliffs, N.J., 1964), pg. 482.
3. Sakharov A.D., *JETP Lett.* **5**, 24 (1967).
4. See any Quantum Field Theory textbook, e.g., Huang K., "*Quarks, Leptons and Gauge Fields*", (World Scientific Pub. Co., Singapore).

5. Klinkhammer F.R. and Manton N.S., *Phys. Rev.* **D30**, 2212 (1984).
6. See an early discussion in Polyakov A., *Sov. Phys. - JETP*, **41**, 988 (1976).
7. Kleihaus B., Kunz J. and Brihaye Y., *Phys. Lett.* **B273**, 100 (1992).
8. Kuzmin V.A., Rubakov V.A., and Shaposhnikov M.E., *Phys. Lett.* **B155**, 36 (1985).
9. Arnold P. and McLerran L., *Phys. Rev.* **D36**, 581 (1987); *Phys. Rev.* **D37**, 1020 (1988).
10. Dine M., Lechtenfeld O., Sakita B., Fischler W., Polchinski J., *Nuc. Phys.* **B342**, 381 (1990).
11. Ambjorn J., Askgaard T., Porter H. and Shaposhnikov M.E., *Phys. Lett.* **B244**, 479 (1990); *Nuc. Phys.* **B353**, 346 (1991).
12. Bochkarev A.I. and Shaposhnikov M.E., *Mod. Phys. Lett.* **A2**, 417 (1987); Bochkarev A.I., Khlebnikov S. Yu. and Shaposhnikov M.E., *Nuc. Phys.* **B329**, 490 (1990).
13. Kirzhnitz D.A. and Linde A.D., *Sov. Phys.-JETP*, **40**, 628 (1974); Dolan J. and Jackiw R., *Phys. Rev.* **D9**, 3320 (1974); Weinberg S., *Phys. Rev.* **D9**, 3357 (1974).
14. Coleman S., *Phys. Rev.* **D15**, 2929 (1977); Callan C. and Coleman S., *Phys. Rev.* **D16**, 1762 (1977).
15. Linde A.D., *Phys. Lett.* **B70**, 306 (1977); *Phys. Lett.* **B100**, 37 (1981); *Nuc. Phys.* **B216**, 421 (1983).
16. Brandenberger R. and A-C. Davis, *Phys. Lett.* **B308**, 79 (1993); Brandenberger R., Davis A-C. and Trodden M., *Phys. Lett.* **B335**, 123 (1994); Brandenberger R. et al., preprint BROWN-HET-962.
17. Jarlskog C., *Phys. Rev. Lett.* **55**, 1039 (1985).
18. Gunion J.F., Haber H.E., Kane G.L. and Dawson S. "*The Higgs Hunters Guide*", (Addison-Wesley 1990).
19. Anderson G.E. and Hall L.J., Phys. Rev. **D45**, 2685 (1992).
20. Cohen A., Kaplan D. and Nelson A., *Ann. Rev. of Nucl. and Particle Science* **43**, 27 (1993).
21. Cohen A., Kaplan D. and Nelson A., *Nuc. Phys.* **B349**, 727 (1991); *Phys. Lett.* **B263**, 86 (1991).
22. McLerran L., Shaposhnikov M.E., Turok N. and Voloshin M. *Phys. Lett.* **B256**, 351 (1991).
23. Glashow S. and Weinberg S., *Phys. Rev.* **D15**, 1958 (1977).
24. Bhowmik Duari S., Ph.D. Thesis, IIT Bombay (1995), unpublished.
25. Dine M., Leigh R.L., Huet P., Linde A. and Linde D., *Phys. Lett.* **B283**, 319 (1992); *Phys. Rev.* **D46**, 550 (1992).

26. Yajnik U.A. *Phys. Rev.* **D34**, 1237 (1986).
27. Bhowmik Duari S. and Yajnik U.A. *Phys. Lett.* **B326**, 212 (1994).
28. Bhowmik Duari S. and Yajnik U.A., *to appear in Nucl. Phys. B, proceedings supplement, Workshop on Astroparticle Physics, Stockholm University 1994.*
29. Yu-Liang-Wu CMU-HEP93-19; DOE-ER/40682-44.
30. Bhowmik Duari S. and Yajnik U.A., to be published. See also ref.24.

Tolman's Solutions in Higher Dimensional Space Time*

T. Singh
Department of Applied Mathematics
Institute of Technology, Banaras Hindu University
Varanasi 221005, India

Summary

In this paper higherdimensional generalization of the class of Tolman's solutions (1939) has been obtained. They are important for a study and understanding of the internal constitution of stars.

1. Introduction

The exact physical situation at very early stages of the formation of our universe provoked great interest among theoretical physicists. Kaluza-Klein view of world geometry is that the universe started out in a higher-dimensional phase with some dimensions eventually collapsing and stabilizing at a size close to the Planck length while three others continued to expand and are still doing so (Sahdev 1984). The explanation of the smallness of the extra dimensions of the universe by the dynamical evolution of the latter has also been proposed in the case of a more realistic model (11-dimensional supergravity). Now in view of recent developments of superstring theory in which the space-time is considered to be of dimension higher than four, the studies in higher dimensions have attained new importance. The generalization of solutions of Einstein's equations to higher dimensions thus becomes a necessity. Myers and Perry (1986) have presented higher dimensional extension of Schwarzschild, Reissner-Nordstrom and Kerr solutions. Xu Dianyan (1988) has obtained the Reissner-Nordstrom-de Sitter metric and Kerr-de Sitter metric for higher dimensional space-time. Krori *et al* (1988, 1989, 1990) have presented higher dimensional generalization of Schwarzschild interior, Floride's and Marder's solutions. Shen and Tan (1989a,b) have obtained higher dimensional generalization of interior Wyman's ($\rho = \mu r^{a}$) solution and global regular solution with equation of state $p + \rho = 0$. Several authors (Koikawa 1986, Iyer and Vishveshwara 1989, Liddle *et al* 1990 and Chatterjee *et al* 1990a,b) have studied different solutions in higher-dimensional space-time.

---

* Dedicated to Prof. V.B. Johri on his sixtieth birthday.

Singh *et al* (1995a,b) have presented the generalization of Mehra's (1966), Whittaker's (1969), Ibrahim-Nutku's (1976) solutions; solutions for superdense, disordered radiation, constant gravitational mass density; Alder (1974), Kuchowicz's (1968, 1975) and Bayin's (1978) solutions to higher dimensional space-time.

The object of this paper is to obtain higher dimensional generalization of the class of Tolman's (1939) solutions which are useful for understanding the internal constitution of stars.

2. The field equations

For higher-dimensional space-time, the Einstein field equations are

$$R_{\mu\nu} = -8\pi G\left(T_{\mu\nu} - \frac{1}{D-2} g_{\mu\nu} T^{\lambda}_{\lambda}\right) \tag{2.1}$$

where D is the number of dimensions taken as $D = n + 3$. The energy momentum tensor is

$$T_{\mu\nu} = \text{diag}\left(\rho, \underbrace{-p, \ldots, -p}_{(n+2)}\right) \tag{2.2}$$

where ρ is the material density and p is the pressure.

We consider the static spherically symmetric line element for D-dimensional space time in the form (Myers and Perry, 1986):

$$\begin{aligned} ds^2 = e^{\nu} dt^2 - e^{\lambda} dr^2 - r^2 (d\theta_1^2 + \sin^2\theta_1\, d\theta_2^2 + \\ \ldots + \sin^2\theta_1 \sin^2\theta_2 \ldots \sin^2\theta_n\, d\theta_{n+1}^2) \end{aligned} \tag{2.3}$$

Here $x^0 = t,\ x^1 = r,\ x^2 = \theta_1,\ x^3 = \theta_2$ etc.

For simplicity, we taken $c = 1$, $G = 1$. Thus the field equation (2.1) for the metric (2.3) leads to set of equations

$$e^{-\lambda}\left(\frac{\lambda'}{r} - \frac{n}{r^2}\right) = \frac{16\pi}{(n+1)}\rho \tag{2.4}$$

$$e^{-\lambda}\left(\frac{\nu'}{r} + \frac{n}{r^2}\right) - \frac{n}{r^2} = \frac{16\pi}{(n+1)}p \tag{2.5}$$

$$e^{-\lambda}\left(\frac{\nu''}{2} + \frac{\nu'^2}{4} - \frac{\lambda'\nu'}{4} - \frac{\nu' + n\lambda'}{2r} - \frac{n}{r^2}\right) + \frac{n}{r^2} = 0 \tag{2.6}$$

$$p' + (p+\rho)\frac{\nu'}{2} = 0 \tag{2.7}$$

Here a prime denotes differentiation with respect to r. Multiplying equation (2.6) by $2/r$ and rearranging the terms, we get

$$\frac{d}{dr}\left[\frac{n(e^{-\lambda}-1)}{r^2}\right] + \frac{d}{dr}\left[\frac{e^{-\lambda}\nu'}{2r}\right] + e^{-\lambda-\nu}\frac{d}{dr}\left[\frac{e^{\nu}\nu'}{2r}\right] = 0 \tag{2.8}$$

It is difficult to find general solution of equation (2.8). Therefore we impose some physically reasonable conditions on λ or ν which makes the equation (2.8) integrable.

3. Specific solutions

Solution I (Einstein Universe in Higher Dimensions)

We assume

$$e^{\nu} = \text{constant} = a \quad \text{(say)} \tag{3.1}$$

With this assumption equation (2.8) reduces to

$$\frac{d}{dr}\left[\frac{n(e^{-\lambda}-1)}{r^2}\right] = 0 \tag{3.2}$$

which on integration gives

$$e^{-\lambda} = 1 - \frac{r^2}{nR^2} \tag{3.3}$$

where $-1/R^2$ is taken as a constant of integration.

Considering the assumption (3.1) from equations (2.5) and (3.3), we have

$$p = \frac{(n+1)}{16\pi n R^2} \tag{3.4}$$

From (2.4) and (3.3),

$$\rho = \frac{(n+2)(n+1)}{16\pi n R^2} \tag{3.5}$$

The resulting solution is static Einstein universe in higher dimension with uniform pressure and density.

Solution II (Schwarzschild Universe)

We choose

$$e^{-\lambda-\nu} = b^2 = \text{constant}. \tag{3.6}$$

In this case equation (2.8) gives the solution

$$e^{\lambda} = \left[1 - \frac{2m}{r^n} - \frac{3r^2}{(n+2)R^2}\right]^{-1} \tag{3.7}$$

where m and R are arbitrary constants.

By use of equations (3.6) and (3.7) pressure and density can be obtained from equations (2.5) and (2.4) respectively as

$$p = \frac{-3(n+1)}{16\pi R^2} \tag{3.8}$$

$$\rho = \frac{3(n+1)}{16\pi R^2} \tag{3.9}$$

Similar solution has been obtained by Shen and Tan (1989b) by choosing equation of state $p + \rho = 0$. We have written this solution only for the sake of completeness. With $R \to \infty$ the solution reduces to higher dimensional analogue of Schwarzschild exterior solution for a particle of mass m.

Solution III (Schwarzschild Interior Solution)

If we take

$$e^{-\lambda} = \left[1 - \frac{r^2}{nR^2}\right] \tag{3.10}$$

the equation (2.8) yields the solution

$$e^{\nu} = \left[d - nCR^2\left(1 - \frac{r^2}{nR^2}\right)^{1/2}\right]^2 \tag{3.11}$$

The pressure and material density can be obtained as

$$p = \frac{(n+1)}{16\pi R^2}\left[(n+2)\,CR^2\left(1 - \frac{r^2}{nR^2}\right)^{1/2} - d\right] \cdot \left[d - nCR^2\left(1 - \frac{r^2}{nR^2}\right)^{1/2}\right]^{-1} \tag{3.12}$$

$$\rho = \frac{(n+1)(n+2)}{16\pi n R^2} \tag{3.13}$$

The solution is higher dimensional generalization of well-known Schwarzschild interior solution for a sphere of fluid with constant density. When $C = 0$ the solution degenerates into higher dimensional Einstein universe as obtained in solution I.

When $d = 0$, we have

$$e^{-\lambda}\left[1 - \frac{r^2}{nR^2}\right] \quad \text{and} \quad e^{\nu} = \text{constant}\left(1 - \frac{r^2}{nR^2}\right) \tag{3.14}$$

which is de-Sitter universe in higher dimensional space-time.

Solution IV

In this case we assume

$$e^{\nu}\frac{\nu'}{2r} = \text{constant} \tag{3.15}$$

which has the solution

$$e^{\nu} = B^2\left[1 + \frac{r^2}{A^2}\right] \tag{3.16}$$

With the assumption (3.15) the equation (2.8) becomes immediately integrable and we get

$$e^{\lambda} = \left[1 + \frac{(n+1)\, r^2}{nA^2}\right] \Big/ \left(1 - \frac{r^2}{nR^2}\right)\left(1 + \frac{r^2}{A^2}\right) \tag{3.17}$$

From equations (2.5) and (3.15)–(3.17), we have

$$\frac{16\pi}{(n+1)}\, p = \frac{1}{A^2}\left[1 - \frac{A^2}{R^2} - \frac{(n+2)\, r^2}{nR^2}\right] \Big/ \left(1 + \frac{(n+1)\, r^2}{nA^2}\right) \tag{3.18}$$

Similarly by use of equation (3.17) from (2.4) one can obtain material density as

$$\frac{16\pi}{(n+1)}\rho=\frac{1}{A^2}\left[1+\frac{(n+2)A^2}{nR^2}+\frac{(n+2)r^2}{nR^2}\right]\Big/\left[1+\frac{(1+n)r^2}{nA^2}\right]$$

$$+\frac{2}{nA^2}\left(1-\frac{r^2}{nR^2}\right)\Big/\left(1+\frac{(1+n)r^2}{nA^2}\right)^2 \tag{3.19}$$

At the centre of sphere the pressure (p_c) and density (ρ_c) have the values

$$\frac{16\pi}{(n+1)}p_c=\frac{1}{A^2}-\frac{1}{R^2}, \tag{3.20}$$

and

$$\frac{16\pi}{(n+1)}\rho_c=\left(1+\frac{2}{n}\right)\left(\frac{1}{A^2}+\frac{1}{R^2}\right), \tag{3.21}$$

The equations (3.18)–(3.21) can be combined in a simple form

$$\rho=\rho_c-(n+4)(p_c-p)-\frac{4(n+1)(p_c-p)^2}{n(\rho_c+p_c)} \tag{3.22}$$

which is 'equation of state' connecting the density and pressure of the fluid inside the sphere.

At the boundary of the sphere the pressure drops to zero. Therefore equation (3.22) with $p=0$ yields

$$\rho_b=\rho_c-(n+4)p_c-\frac{4(n+1)p_c^2}{n(\rho_c+p_c)} \tag{3.23a}$$

The equation (3.18) suggests that at the boundary of the sphere

$$r_b=R\left[\frac{n}{(n+2)}\left(1-\frac{A^2}{R^2}\right)\right]^{1/2} \tag{3.23b}$$

It is clear that with $R^2>A^2$, the pressure and density of the fluid fall from their central to their boundary values where the density still remains positive.

The solution may be useful in studying properties of spheres of compressible fluid in higher dimensions since the equation of state (3.22) is relatively simple.

Solution V

If we consider

$$e^{\nu} = B^2 r^{2a}, \; a, \; b = \text{const.}; \tag{3.24}$$

then equation (2.8) gives the solution

$$e^{\lambda} = \frac{n + 2a - a^2}{n - (n + 2a - a^2)(r/R)^N} \tag{3.25}$$

where

$$N = \frac{2(n + 2a - a^2)}{(n + a)} \tag{3.26}$$

With the help of equations (3.24) and (3.25) from equations (2.4) and (2.5) one can obtain the expression for density and pressure as

$$\rho = \frac{(n+1)}{16\pi}\left[\frac{n(2a - a^2)}{(n + 2a - a^2)r^2} + \frac{(n + N)}{R^2}\left(\frac{r}{R}\right)^{N-2}\right] \tag{3.27}$$

$$p = \frac{(n+1)}{16\pi}\left[\frac{a^2}{(n + 2a - a^2)r^2} - \frac{(n + 2a)}{R^2}\left(\frac{r}{R}\right)^{N-2}\right] \tag{3.28}$$

This solution corresponds to a higher dimensional fluid sphere of infinite density and pressure at the centre. The pressure and density both decrease from an infinite value at the centre to zero pressure at the boundary where density still remains positive.

Solution VI

Firstly we will establish higher dimensional analogue of the general solution of Wyman (1949) and then obtain analogue of Tolman's solution as a special case.

We shall take

$$e^{\nu} = (A\, r^{1-a} - B\, r^{1+a})^2 \tag{3.29}$$

where a, A and B are constants.

Using (3.29) in equation (2.8) we obtain

$$\frac{d}{dr}(e^{-\lambda}) + \frac{2\,(a^2 - n - 1)\,A - B\,r^{2a}}{[(1+n-a)\,A - (1+n+a)\,Br^{2a}]\,r}\,e^{-\lambda}$$

$$+ \frac{2n\,[A - B\,r^{2a}]}{[(1+n-a)\,A - (1+n+a)\,Br^{2a}]\,r} = 0 \qquad (3.30)$$

which is a linear equation of first order in $e^{-\lambda}$. Its solution can be obtained as

$$e^{-\lambda} = \frac{n}{(1+n-a^2)} + dr^b\,[(1+n-a)\,A - (1+n+a)\,B\,r^{2a}]^c \qquad (3.31)$$

where d is an arbitrary constant of integration and

$$b = \frac{2\,(a^2 - n - 1)}{(a - n - 1)}, \quad c = \frac{2\,(1+n-a^2)}{[a^2 - (1+n)^2]} \qquad (3.32)$$

This solution is valid provided a does not have the value $(n+1)$ or $(n+1)^{\frac{1}{2}}$.

By use of equation (3.31) into equation (2.4) we obtain

$$\rho = \frac{n\,(n+1)\,(1-a^2)}{16\,(1+n-a^2)\,\pi\,r^2} - \frac{d\,(n+1)}{16\pi}\,[n\,(1+n-a) + 2\,(1+n+a)]$$

$$[A - Br^{2a}]\,[(1+n-a)\,A - (1+n+a)\,B\,r^{2a}]^{c-1}\,r^{b-2} \qquad (3.33)$$

Similarly from equations (2.5), (3.29) and (3.31) one can obtain the expression for pressure as

$$p = \frac{n\,(n+1)\,[(1-a)^2\,A - (1+a)^2\,B\,r^{2a}]}{16\pi\,(1+n-a^2)\,(A - B\,r^{2a})\,r^2}$$

$$+ \frac{d\,(n+1)\,[(1+n-a)\,A - (1+n+a)\,Br^{2a}\,(2+n-2a)\,A - (2+n+2a)\,Br^{2a}]}{16\pi\,r^{2-b}\,(A - Br^{2a})} \qquad (3.34)$$

This solution is a higher dimensional generalization of Wyman's solution. If we take $d = 0$ it gives Tolman's solution VI in higher dimensional space-time.

Solution VII

For seventh solution in the sequence, we assume

$$e^{-\lambda} = 1 - \frac{r^2}{R^2} + \frac{4r^4}{A^4} \qquad (3.35)$$

With this assumption equation (2.8) becomes so complicated that the solution is not a convenient one for physical considerations. Therefore we shall not pursue its study.

Solution VIII

Now we assume

$$e^{-\lambda} = B^{-2} r^{-2b} e^{\nu} \tag{3.36}$$

where b and B are arbitrary constants. With this assumption equation (2.8) reduces to

$$r^2 \frac{d^2}{dr^2}(e^{\nu}) + (n - b - 1)\, r \frac{d}{dr}(e^{\nu}) - 2n\,(b + 1)\, e^{\nu}$$

$$= -2n\, B^2\, r^{2b} \tag{3.37}$$

This is a second order non-homogeneous linear differential equation. Its solution can be obtained in the form

$$e^{\nu} = C_1\, r^{D_1} + C_2\, r^{D_2} + \frac{n\, B^2\, r^{2b}}{(n + 2b - b^2)} \tag{3.38}$$

where C_1, C_2 are constants and D_1, D_2 are defined as

$$D_1 = \frac{(1 + b - n) + \sqrt{(n - b - 1)^2 + 8n\,(b + 1)}}{2},$$

$$D_2 = \frac{(1 + b - n) - \sqrt{(n - b - 1)^2 + 8n\,(b + 1)}}{2};$$

Using equations (3.36) and (3.38) from equations (2.4) and (2.5) one can get the value for ρ and p as

$$\frac{16\pi}{(n + 1)}\,\rho = \frac{1}{B^2\, r^{2b + 2}}\left[(2b - D_1 - n)\, C_1\, r^{D_1} + (2b - D_2 - n)\, C_2\, r^{D_2}\right]$$

$$+ \frac{nb\,(2 - b)}{(n + 2b - b^2)\, r^2} \tag{3.39}$$

$$\frac{16\pi}{(n+1)}p = \frac{1}{B^2 r^{2b+2}}\left[(n+D_1)C_1 r^{D_1} + (n+D_2)C_2 r^{D_2}\right] + \frac{nb^2}{(n+2b-b^2)r^2} \tag{3.40}$$

From equations (3.38)-(3.40) we have

$$\frac{16\pi}{(n+1)}(p+\rho) = \frac{2b}{B^2 r^{2b+2}}e^{\nu} \tag{3.41}$$

As a particular case if we take $b = 1 + \sqrt{1+n}$, the equation (3.37) has the solution

$$e^{\nu} = \alpha r^{2b} + \beta r^{(2-n-b)} - \frac{2n B^2 r^{2b}\log r}{(3b+n-2)}, \tag{3.42}$$

where α and β are constants.

In this case energy density and pressure become

$$\frac{16\pi}{(n+1)}\rho = \frac{n(B^2-\alpha)}{B^2 r^2} + \frac{(3b-2)\beta}{B^2 r^{n+3b}} + \frac{2n(1+n\log r)}{(3b+n-2)r^2} \tag{3.43}$$

$$\frac{16\pi}{(n+1)}p = \frac{2b\alpha + n\alpha - nB^2}{B^2 r^2} + \frac{(2-b)\beta}{B^2 r^{n+3b}} + \frac{2n(1+(2b+n)\log r)}{(3b+n-2)r^2} \tag{3.44}$$

This solution is a higher dimensional generalization of Wyman's (1949) solution.

These solutions can be matched at the boundary $r = r_b$ with the exterior solution (Myers and Perry 1986):

$$ds^2 = \left(1 - \frac{C}{r^{D-3}}\right)dt^2 - \left(1 - \frac{C}{r^{D-3}}\right)^{-1} dr^2 - r^2(d\theta_1^2 + \sin^2\theta_1\, d\theta_2^2 + \ldots + \sin^2\theta_1 \ldots \sin^2\theta_n\, d\theta_{n+1}^2) \tag{3.45}$$

where C is the total mass of the fluid inside a sphere of radius r_b given by

$$M = \frac{1}{2}C(D-2)A_{D-2} \tag{3.46}$$

where

$$A_{D-2} = 2\pi^{(D-2)/2}(D-1)/2. \tag{3.47}$$

References

1. Alder, R.J. (1974) J. Math. Phys., **15**, 727.
2. Bayin, S.S. (1978) Phys. Rev. D, **18**, 2745.
3. Chatterjee, S., Banerjee, A. and Bhui, B. (1990a). Phys. Lett., A, **149**, 91.
4. Chatterjee, S., Bhui, B. and Banerjee, A. (1990b). J. Math. Phys., **31**, 2203.
5. Florides, P.S. (1974) Proc. Roy. Soc. A, **337**, 592.
6. Ibrahim, A. and Nutku, Y. (1970). Gen. Rel. Grav., **7**, 949.
7. Iyer, B.R. and Vishveshwara, C.V. (1989). Pramana (India), **32**, 749.
8. Koikawa, T. (1986). Phys. Lett. A., **117**, 729.
9. Krori, K.D., Borgohain, P. and Das, K. (1988). Phys. Lett., A, **132**, 321.
10. Krori, K.D., Borgohain, P. and Das, K. (1989). Gen. Rel. Grav., **21**, 1099.
11. Krori, K.D., Borgohain, P. and Das, K. (1990). J. Math. Phys., **31**, 149.
12. Kuchowicz, B. (1968). Acta. Phys. Polon B., **33**, 541.
13. Kuchowicz, B. (1975). Astrophys. Space Sci., **33**, L13.
14. Liddle, A.R., Moorhouse, R.G. and Henriques, A.B. (1990) Class Quant. Grav. (U.K.), **7**, 1009.
15. Marder, L. (1958). Proc. R. Soc. London Ser. A, **244**, 524.
16. Mehra, A.L. (1966). J. Austr. Math. Soc., **6**, 153.
17. Myers, R.C. and Perry, M.J. (1986). Ann. Phys. (NY), **172**, 394.
18. Sahdev, D. (1984). Phys. Rev. D, **30**, 2495.
19. Shen, Y.G. and Tan, Z.D. (1988a). Phys. Lett. A, **137**, 96.
20. Shen, Y.G. and Tan, Z.D. (1989b). Phys. Lett. A, **142**, 341.
21. Singh, T., Singh, G.P. and Helmi, A.M. (1995a). Some Fluid Spheres in Higher Dimensional Space-Time (Communicated for publication).
22. Singh, T., Singh, G.P. and Helmi, A.M. (1995b). Static solutions of Einstein's Field equations for Fluid Spheres in higher - dimensions (Communicated for publication).
23. Tolman, R.C. (1939). Phys. Rev., **55**, 364.
24. Whittaker, J.M. (1968). Proc. Roy. Soc. **A306**, 1.
25. Wyman, M. (1949). Phys. Rev., **75**, 1930.
26. Xu Dianyan (1988). Class Quant. Grav. (U.K.) **5**, 871.

On the Cosmological Constant Problem

S.K. Srivastava
Department of Mathematics
North Eastern Hill University, Bijni Complex
Bhagyakul, Shillong 793 003. (India).

Abstract

Five-dimensional cosmological model driven by scalar fields is derived. In this model, time-dependent cosmological constant as well as effective gravitational constant, induced through dimensional reduction and one-loop quantum correction, has been discussed. It is interesting to see that $\Lambda_{eff} \to \infty$ as $t \to 0$ and $\Lambda_{eff} \to 0$ as $t \to \infty$. Also $G_{eff} \to G_N$ (Newtonian gravitational constant) when $t \to \infty$ and $G_{eff} \to 0$ as $t \to 0$.

After formulation of general theory of relativity which provided an effective theory of gravity at low energy, Einstein attempted to apply it to the whole universe. He assumed the universe to be static. So to achieve a static solution, he added a new term involving a free parameter Λ, called cosmological constant, to his field equations. After Hubble's discovery of expansion of the universe, Einstein had to retract his proposal of adding cosmological term. But, unfortunately it could not be dropped out, because anything contributing to vacuum energy density acts like cosmological constant. It is expected that today energy density due to the cosmological constant is extremely small, even less then $10^{-48}\,\text{GeV}^4$. But if one believes general relativity at Planck scale, one gets its value $\sim 10^{76}\,\text{GeV}^4$. The problem is to understand how the value of energy density due to cosmological constant comes down to 119 decimal places today. In this article, a model is suggested which results into time-dependent cosmological constant providing a possible solution to this problem.

This article is planned as follows. In the beginning Einstein's equations are solved for 5-dim. Robertson-Walker type space-time followed by a discussion on cosmological compactification of the model. After this, dimensionally reduced 4-dim. action for gravity is obtained from 5-dim. action for gravity for the cosmological model under consideration. After this, one-loop quantum correction for the scalar field is discussed which contributes to effective gravity and cosmological term. Results obtained for time-dependent cosmological constant and gravitational constant is, then, discussed in the context of solutions of Einstein's equations. $\hbar = c = 1$ is used, throughout the entire article, as fundamental unit where $\hbar$ and c have their usual meaning.

In the context of unification of gravity with other fundamental forces of the nature, relativistic dynamics in higher dimensional spaces (spaces with dimension more than four) is important. Here 5-dimensional space-time, which was originally proposed by Kaluza-Klein in the context of unification of gravity and electromagnetism[1], is considered. The topology of the space-time, under consideration, here is $\mathbf{R}$ (time) $\otimes M^3$ (space) $\otimes S^1$ (where S^1 is circle). x^μ are coordinates for IR $\otimes M^3$ and y denotes the coordinate on S^1. The line-element is the generalized Robertson-Walker line-element for (4 + 1) – dimensional space-time which is given as

$$ds^2 = dt^2 - r^2(t) \sum_{i,j=1}^{3} \delta_{ij}\, dx^i\, dx^j - R^2(t)\, dy^2 \tag{1}$$

where $r(t)$ and $R(t)$ are scale factors.

A massless and minimally coupled scalar $\phi\,(x, y)$ is taken to be the matter field having the action given as

$$S_\phi = \frac{1}{2} \int d^4x\, dy \sqrt{-g_5}\, g^{mn}\, \partial_m \phi^*\, \partial_n \phi \tag{2}$$

where g^{mn} is inverse of the metric tensor g_{mn} (m, n = 0, 1, 2, 3, 4) of the space-time (1) and g_5 is the determinant of g_{mn}. The energy momentum tensor is calculated from (2) as

$$T_{mn} = \partial_m \phi^*\, \partial_n \phi - \frac{1}{2}\, g_{mn} (g^{m'n'}\, \partial_{m'} \phi^*\, \partial_{n'} \phi) \tag{3}$$

Further, $\phi(t, \vec{x}\,; y)$ is decomposed as

$$\phi(t, \vec{x}\,; y) = \phi_0(t, 0; 0) + \bar{\phi}\,(t, \vec{x}\,; y) \tag{4}$$

where $\phi_0(t)$ is homogeneous and $\bar{\phi}$ is the non-homogeneous solution of Klein-Gordon equation obtained from the action (2). Also it is assumed that $\bar{\phi}$ is small fluctuation, so the cosmic dynamics is mainly governed by ϕ_0. It is good to mention here that though role of $\bar{\phi}$ is not significant in cosmic dynamics but it is important otherwise.

Having these assumptions, in mind, one can calculate, the Einstein's field equations for the space-time (1) as

$$3\frac{\ddot{r}}{r}+\frac{\ddot{R}}{R}=-8\pi G_5\phi_0^2(t) \tag{5a}$$

$$\frac{d}{d\tilde{t}}\left(\frac{\dot{r}}{r}\right)+\frac{\dot{r}}{r}\left(3\frac{\dot{r}}{r}+\frac{\dot{R}}{R}\right)=0 \tag{5b}$$

$$\frac{d}{d\tilde{t}}\left(\frac{\dot{R}}{R}\right)+\frac{\dot{R}}{R}\left(3\frac{\dot{r}}{r}+\frac{\dot{R}}{R}\right)=0 \tag{5c}$$

Here dot (·) denotes derivative with respect to a dimensionless parameter $\tilde{t}=t/t_p$ (t is the cosmic time and t_p is the Planck time). $G_5=G_N\,L$ ($0\leq y\leq L$ and G_N is the Newtonian gravitational constant which is equal to (Planck Mass)$^{-2}$). Writing

$$V=r^3R \tag{6}$$

One finds

$$3\frac{\dot{r}}{r}+\frac{\dot{R}}{R}=\frac{\dot{V}}{V} \tag{7}$$

Moreover, as mentioned above, $\phi_0(t)$ satisfies the Klein-Gordon equation

$$0=\Box_5\,\phi_0=\frac{1}{\sqrt{-g_5}}\frac{\partial}{\partial x^m}\left(\sqrt{-g_5}\,g^{mn}\frac{\partial}{\partial x^n}\right)\phi_0=-\frac{1}{V}\frac{d}{dt}\left(V\frac{d\phi}{dt}\right) \tag{8}$$

which yields the solution

$$\frac{d\phi}{dt}=\frac{\alpha}{V} \tag{9}$$

where α is constant of integration.

Also (5b) yields on integration

$$\frac{\dot{r}}{r}=\frac{1}{V} \tag{10}$$

absorbing the constant of integration in V.

Connecting (5), (7), (9) and (10), after a small manipulation, one gets (absorbing α in L as $G_5=G_NL$)

$$6\dot{V}=12+8\pi G_5 \tag{11}$$

which integrates to

$$6V = (12 + 8\pi G_5)\,\tilde{t} \tag{12}$$

with initial condition $V(0) = 0$.

Now (10) and (12) yield the solution

$$r = r_0\,(\tilde{t})^{6/12 + 8\pi G_5} \tag{13}$$

where r_0 is an integration constant. Introducing $r(\tilde{t})$ from (13) into (6), one gets

$$R(\tilde{t}) = \left(\frac{12 + 8\pi G_5}{6r_0^3}\right)(\tilde{t})^{\frac{8\pi G_5 - 6}{8\pi G_5 + 12}} \tag{14}$$

(13) and (14) show that $r(t)$ expands and $R(t)$ contracts with time if $L < (6M_p^2)/(8\pi)$ (M_p is the Planck mass). Our observable universe is 4-dimensional. So, if one assumes the higher-dimensional universe, the extra (internal) manifold (for example S^1 in this case) should compactify to an undetectable size (extremely small size). It cannot be possible if $L \geq (6M_p^2)/(8\pi)$. Hence on the basis of this argument, the only possibility remains that $L << (6M_p^2)/(8\pi)$. So, (14) shows that $R(t)$ contracts with time and has a "crack of doom singularity" as $t \to \infty$.

The effective radius of the internal manifold, S^1, is $\rho\, R(t)$ where ρ is the physical radius such that $L = 2\pi\rho$. Using (14), one gets

$$\rho R(t) = \left(\frac{12 + 8\pi G_5}{6r_0^3}\right)\rho\,(\tilde{t})^{\frac{8\pi G_5 - 6}{8\pi G_5 + 12}} \tag{15}$$

Since ρ and r_0 both are arbitrary, one can safely assume that

$$\rho\,(12 + 8\pi G_5) = 6r_0^3\, L_p \tag{16}$$

where L_p is the Planck length.

Under this assumption, the effective radius is equal to

$$L_p\,(\tilde{t})^{\frac{8\pi G_5 - 6}{8\pi G_5 + 12}} \tag{17}$$

which shows that the effective radius was extremely large for small t and extremely small for large t. If extra manifold is actually hidden, the effective radius should not be greater than L_p at the compactification time, t_c (the time when the universe is supposed to be approximately 4-dim.). Looking at (17), one finds the effective radius equal to l_* at $\tilde{t}_c = 1$ or $t_c = t_p$.

The 5-dim. action for gravity, in general, is written as

$$S_g^{(5)} = -\frac{1}{16\pi G_5}\int d^4x\, dy \sqrt{-g_5}\, R_5 \tag{18}$$

where R_5 is the 5-dimensional Ricci scalar. The action given by (18) can be dimensionally reduced to 4-dim. action for gravity through conformal transformations[2]. In this method, g_{mn} is transformed to g'_{mn} as

$$g_{mn} = R^2(t) g'_{mn} = R^2(t)\begin{pmatrix} \tilde{g}_{\mu\nu} & 0 \\ 0 & -1 \end{pmatrix} \tag{19}$$

where $\tilde{g}_{\mu\nu}$ ($\mu, \nu = 0,1,2,3$) is the resulting metric tensor on 4-dim. external manifold. So, ignoring term of total divergence, (18) can be re-written as

$$S_g^{(5)} = -\frac{1}{16\pi G_5}\int d^4x\, dy \sqrt{-\tilde{g}_4}\, R^3 \,[\tilde{R}_4 - 12R^{-2}(\tilde{\nabla} R)^2] \tag{20}$$

where $\tilde{\nabla}$ is the covariant derivative and $\tilde{R}_4$ is the Ricci scalar with respect to $\tilde{g}_{\mu\nu}$.

Further conformal transformation is done over $\tilde{g}_{\mu\nu}$ only as

$$\tilde{g}_{\mu\nu} = e^{2v} g_{\mu\nu} \tag{21}$$

where v is function of $R(t)$. Now using (21) and integrating over y, one gets from (20)

$$S_g^{(4)} = -\frac{1}{16\pi G_N}\int d^4x \sqrt{-g_4}\, R^3 e^{2v}\left[R_4 - 12\left(\frac{dv}{dt}\right)^2 - 12\left(\frac{1}{R}\frac{dR}{dt}\right)^2 - \frac{8}{R}\frac{dR}{dt}\frac{dv}{dt}\right] \tag{22}$$

choosing $v = -\frac{3}{2}\ln R$, one gets dimensionally reduced 4-dim. action for gravity as

$$S_g^{(4)} = -\frac{1}{16\pi G_N}\int d^4x \sqrt{-g_4}\left[R_4 - 12\left(\frac{1}{R}\frac{dR}{dt}\right)^2\right] \tag{23}$$

Introducing decomposition (4) into the action (2) for the ϕ – field, one gets

$$S_\phi = S[\phi_0(t)] + S_2[\bar{\Phi}(t, \vec{x}; y)] \tag{24}$$

In Taylor's expansion (24), first order term is absent because $\left.\frac{\delta S}{\delta \phi}\right|_{\phi=\phi_0} = \Box_5 \phi_0 = 0$. The reason for absence of higher order terms is that higher order derivatives of action with respect to ϕ are vanishing.

The extra manifold is a circle which is not simply connected, hence my field on it can be untwisted (periodic in y) or twisted (anti-periodic in y)[3]. Hence in either case, one may write a $\bar{\phi}(t, \vec{x}; y)$ as

$$\bar{\phi}(t, \vec{x}; y) = (RL)^{-1/2} \sum_{n=-\infty}^{\infty} \bar{\phi}_n(t, \vec{x}) \exp[i(n+\alpha)My] \qquad (25)$$

where $M = 2\pi L^{-1}$ and $\alpha = 0\left(\frac{1}{2}\right)$ for untwisted (twisted) fields[4]. Here M is the compactification mass. If one believes in the inequality $L << \frac{6M_p^2}{8\pi}$, introducing $L = 2\pi M^{-1}$, one finds a consistency check that $M >> \frac{8\pi^2}{3} M_p^{-2}$, which is reasonable as M is supposed to be of the order of Planck mass.

Introducing $\bar{\phi}$ given by Anstatz (25) into $S_2[\bar{\phi}]$ (the second order term in the Taylor's expansion of S_ϕ) and integrating over y, one gets sum of dimensionally reduced 4-dim. action for the scalar fields over all modes as

$$S_2^{(4)}[\bar{\phi}] = \frac{1}{2} \sum_{n=-\infty}^{\infty} \int d^4x \sqrt{-g_4}\, [g^{\mu\nu} (\partial_\mu \bar{\phi}_n)^* (\partial_\nu \bar{\phi}_n) - M_n^2 \bar{\phi}_n^* \bar{\phi}_n] \qquad (26)$$

where

$$M_n^2 = \frac{(n+\alpha)^2 M^2}{R^2} - \frac{3}{2rR} \frac{dr}{dt} \frac{dR}{dt} - \frac{1}{4}\left(\frac{1}{R}\frac{dR}{dt}\right)^2 - \frac{1}{2}\frac{d}{dt}\left(\frac{1}{R}\frac{dR}{dt}\right) \qquad (27)$$

Now one loop effective action is calculated for the nth mode of 4-dim. scalar field ϕ_n and summed up over all modes to get

$$\Gamma_{\bar{\phi}}^{(1)} = \frac{i}{2} \sum_{n=-\infty}^{\infty} \ln \det \Delta_n \qquad (28)$$

where Δ_n is the operator defined, as

$$\Delta^n = \Box_4 + M_n^2 \qquad (29)$$

with

$$\Box_4 = \frac{1}{\sqrt{-g_4}} \frac{\partial}{\partial x^4}\left(\sqrt{-g^4}\, g^{\mu\nu} \frac{\partial}{\partial x^\nu}\right)$$

Using the kernel $K_n(s, x, x)$ for Δ_n, (28) can be re-written as

$$\Gamma_\phi^{(1)} = -\frac{i}{2} \sum_{n=-\infty}^{\infty} \int d^4x \sqrt{-g_4} \int_0^\infty \frac{ds}{s}\, tr\, K_n(s, x, x) \tag{30}$$

where

$$K_n(s, x, x) = i\mu^{4-N} (4\pi i s)^{-\frac{N}{2}} \exp(-iM_n^2 s) \sum_{k=0}^{\infty} (is)^k a_k(x) \tag{31}$$

In (31) N is the space-time dimension used for dimensional regularization with $N \to 4$ at the end of calculation and μ is a constant of mass dimension to get dimensionless action. For Δ_n given by (29)[5,6]

$$a_0(x) = 1 \tag{32a}$$

$$a_1(x) = \frac{1}{6} R_4 \tag{32b}$$

other coefficients a_2, a_3, ... are not mentioned here, because these are not relevant for getting cosmological constant as well as gravitational constant. So, for the purpose of this article a_0 an a_1 are enough. Thus, here complete one-loop effective action is not needed.

Connecting (30), (31) and (32) and integrating over s, one gets

$$\Gamma_\phi^{(1)} = -\frac{1}{2(4\pi)^2} \int d^4x \sqrt{-g_4} \left[\lim_{N\to 4} \Gamma\!\left(-\frac{N}{2}\right) \sum_{n=-\infty}^{\infty} \left\{ \frac{(n+\alpha)^2 M^2}{R^2} + M^{-2} \right\}^{N/2} \right.$$

$$\left. + \lim_{N\to 4} \frac{1}{6} \Gamma\!\left(1 - \frac{N}{2}\right) \sum_{n=-\infty}^{\infty} \left\{ \frac{(n+\alpha)^2 M^2}{R^2} + M^{-2} \right\}^{\frac{N}{2}-1} R_4 + \ldots \right] \tag{33}$$

where

$$\overline{M}^2 = -\frac{3}{2rR}\frac{dr}{dt}\frac{dR}{dt} - \frac{1}{4}\left(\frac{1}{R}\frac{dR}{dt}\right)^2 - \frac{1}{2}\frac{d}{dt}\left(\frac{1}{R}\frac{dR}{dt}\right) \text{ and}$$

Γs is the Gamma function.

Introducting solutions (13) and (14) for the cosmological model, considered here,

$$\overline{M}^2 = -\frac{17}{2}\frac{(8\pi G_5 - 6)}{(8\pi G_5 + 12)^2}\left[1 + \frac{1}{34}(8\pi G_5 - 6)\right]\overline{t}^2$$

In case, $L << \frac{6M_p^2}{8\pi}$ **(which is needed for compactification),** $8\pi G_5 << 6$**. So,**

$$\overline{M}^2 \approx \frac{7}{54t^2} \tag{34}$$

Using formulae (B6) of ref. (7)

$$\sum_{n=-\infty}^{\infty}\left[(n+c)^2 + d^2\right]^{-\lambda} = \pi^{1/2} d^{1-2\lambda}\frac{\Gamma\left(\lambda - \frac{1}{2}\right)}{\Gamma\lambda} + 4\sin\pi\lambda\, f_\lambda(c,d) \tag{35}$$

(where $\mathrm{Re}\lambda > \frac{1}{2}$ and c & d are real numbers), series in (33) is summed up to yield

$$\Gamma_\phi^{(1)} = -\frac{1}{2(4\pi)^2}\int d^4x\sqrt{-g_4}\left[-\frac{8\pi}{15}(\overline{M})^5\frac{R}{M} + \frac{2\pi}{9}\frac{(\overline{M})^3 R}{M}R^4 + \ldots\right] \tag{36}$$

If the number of twisted and untwisted scalar fields is N_0,

$$\Gamma_\phi^{(1)} = -\frac{N_0}{2(4\pi)^2}\int d^4x\sqrt{-g_4}\left[-\frac{8\pi}{15}(\overline{M})^5\frac{R}{M} + \frac{2\pi}{9}\frac{(\overline{M})^3 R}{M}R^4 + \ldots\right] \tag{37}$$

Adding (23) and (37) one gets 4-dim. action for effective gravity as

$$S_g^{(4)\,eff} = \int d^4x\sqrt{-g_4}\left[\left\{\frac{3M_p^2}{4\pi}\left(\frac{1}{R}\frac{dR}{dt}\right)^2 + \frac{N_0}{60\pi}\frac{R}{M}(\overline{M})^5\right\}\right. \tag{38}$$

$$\left. - \left\{\frac{1}{16\pi G_N} + \frac{N_0(\overline{M})^3}{144\pi M}R\right\}R_4 + \ldots\right]$$

(38) shows that

$$\frac{1}{16\pi G_{eff}} = \frac{1}{16\pi G_N} + \frac{N_0(\overline{M})^3 R}{144\pi M} \tag{39a}$$

and

$$\frac{\Lambda_{eff}}{8\pi G_{eff}} = \frac{3M_p^2}{4\pi}\left(\frac{1}{R}\frac{dR}{dt}\right)^2 + \frac{N_0}{60\pi}\frac{R}{M}(\overline{M})^5 \tag{39b}$$

Results (39) can be analyzed in the context of the 5-dim cosmological model considered in the beginning.

Keeping $L << \frac{6M_p^2}{8\pi}$, in mind, R can be approximated as

$$R(\widetilde{t}) \simeq L_p(\widetilde{t})^{-1/3} \tag{40}$$

using (16). As discussed above, compactification time t_c is equal to t_p so compactifiction mass $M = M_p = L_p^{-1}$, in natural units used here. So, using (34), on gets from (39a), 4-dim. effective gravitational constant G_{eff} as

$$\frac{1}{G_{eff}} \simeq \frac{1}{G_N}\left(1 + \left(\frac{7}{54}\right)^{3/2}\frac{N_0}{M_p^{11/3}}\,\overline{t}^{-10/3}\right) \tag{41}$$

From (39b), vacuum energy density

$$\frac{\Lambda_{eff}}{8\pi G_{eff}} = \frac{M_p^2}{12\pi t^2} + \frac{N_0}{60\pi M_p^{5/3}}\left(\frac{7}{54}\right)^{5/2}\frac{1}{t^5} \tag{42}$$

and

$$\Lambda_{eff} \approx \frac{1}{12\pi t^2} \tag{43}$$

Eq.(41) shows that when $t \to 0$, $G_{eff} \to 0$ and when $t \to \infty$, $G_{eff} \to G_N$. Energy density due to cosmological constant (vacuum energy density), obtained from (42), is approximately given as

$$\frac{\Lambda_{eff}}{8\pi G_{eff}} \approx \begin{cases} \frac{N_0}{60\pi}\left(\frac{7}{54}\right)^{5/2}\frac{t^{-5}}{M_p^{5/3}} & \text{for } t << t_p \\ \frac{M_p^2}{12\pi t^2} & \text{for } t \geq t_p \end{cases} \tag{44}$$

Thus, one finds that at Planck time which is compactification time too for the model considered here,

$$\frac{\Lambda_{eff}}{8\pi G_{eff}} \simeq 10^{76}\, GeV^4$$

and taking the present age of the universe around 10^{17} secs. which is equal to

$$\sim \frac{10^{42}}{6.58}\, GeV^{-1}\, (1\, GeV^{-1} = 6.58 \times 10^{-25}\, \text{sec.})$$

$$\frac{\Lambda_{eff}}{8\pi G_{eff}} \simeq 10^{-46}\, GeV^4$$

The interesting point of the result (42) is that it is manifestation of internal manifold in the 4-dim. theory and explains how vacuum energy density comes down with time from extremely large value to extremely small value at present[8].

References

1. Th. Kaluza, Sitzungber; Preuss. Akad. Wiss. Berlin, Math. Phys. Kl, 966 (1921); O. Klein, Z. Phys. **37** (1926) 89; T. Appelquist, A. Chodos and P.G.O. Freund "Modern Kaluza-Klein Theories", (Addison - Wesley, Pub. Comp. Inc. 1987).
2. M.D. Pollock, ICTP preprint IC/90/39.
3. C.J. Isham, Proc. R. Soc. (Lond.) A, **362**, 383 (1978).
4. D.J. Toms, Phys. Lett. B, **129**, 31 (1983).
5. B.S. DeWitt, 'The dynamical theory of groups and fields' in 'Relativity, groups and topology' eds. B.S. DeWitt and C. DeWitt (New York, Gordon & Breach), 1965; Phys. Rep. **19c**, 297 (1975).
6. P.B. Gilkey, J. Diff. Geo. **10**, 601 (1975).
7. L.H. Ford, Phys. Rev. **D 21**, 933 (1980).
8. S.K. Srivastava, J. Math. Phys. **33** 3117 (1992).

Gravitation, New Interactions and Dark Matter : The Emerging Experimental Front

C.S. Unnikrishnan
Gravitation Experiments Group,
Tata Institute of Fundamental Research,
Homi Bhabha Road, Bombay - 400 005, India.

Introduction

In this talk I am attempting to sketch some fragmentary glimpses of experimental explorations in gravitation and related fields which have direct relevance to physics of the early Universe. Study of the evolution of the universe provides a very effective test bench for theoretical hypothesis concerning fundamental particles and interactions. This evolution is governed by gravitation, through the Einstein's equations, and by the various fundamental interactions the standard model of particle physics is attempting to describe. Therefore the early universe is considered to be a laboratory for testing several of our theoretical hypothesis, within or beyond the standard model. The experiments and observations I touch upon in this talk has some resemblance to this situation in that these experiments probe aspects of gravitation as well as of other (new) interactions beyond the standard model.

Methods of understanding and formulating fundamental physical laws have evolved considerably over the past several decades. The most noticeable paradigm shift is the greater reliance placed on invariance principles and on considerations of symmetry in developing theories of physical phenomena. Considerable progress has been made and there is a working theory - the standard model. Though most presently known phenomena are well explained within this model, its predictive power is grossly inadequate on issues like unification of fundamental forces or the existence of new interactions and particles.

Attempts to go beyond the standard model have resulted in many theories, most of which are based on considerations of possible symmetries of nature and on requirements of consistency, renormalizability etc. There are experimentally accessible predictions from these theories. With the natural emergence of new particles in such theories, there are also possible new interactions, mediated by these particles. Typically, these interactions are expected to have coupling strengths comparable to that of gravity and therefore it is conceivable that precision experiments which probe gravitation might get a glimpse of new phenomena. The importance of such findings cannot be overemphasized.

We also have a standard model of cosmology and evolution of the universe with the core structure described by the big bang hypothesis. This model in its simplest form is known to be highly successful in some aspects and grossly inadequate in some other aspects and the need to go beyond this model is clearly dictated by many observations. The most remarkable of these observations is the fact that most of the mass density of the universe is in some non luminous form of matter - the dark matter. This issue is closely linked to the missing links in particle physics.

The experiments I am going to discuss has implications in the study of gravitation, new interactions and dark matter in the universe. These experiments, most of them employing torsion balances as the main detector, are among the most important non-accelerator experiments which have relevance to particle physics and cosmology.

Gravitation and new feeble interactions

The principle of equivalence was the fundamental guiding principle in the formulation of the General Theory of Relativity. Though the empirical statement of the principle was known from Galileo's time, the deep implications had to wait till Einstein's realization that the principle implies the local equivalence of a gravitational field and a nonrotating accelerated frame. In its simplest form the Weak Equivalence Principle states that in a gravitational field all test particles of the same initial velocity fall with same acceleration, independent of composition or structure (universality of free fall - UFF). The Einstein Equivalence Principle (EEP) states that in locally freely falling frames, called Lorentz frames, the non-gravitational laws of physics are those of special relativity. The WEP acquires deep significance within metric theories of gravity because of a conjecture due to L.I. Schiff which states that any complete and self consistent theory of gravity that embodies WEP necessarily embodies EEP and therefore is a metric theory[1].

The WEP is the statement that the inertial mass bears the same ratio to the gravitational mass for all materials independent of their composition and structure. It was Baron von Eötvös who provided firm empirical evidence for a possible true equality of inertial and gravitational mass by setting any possible violation of the WEP at less than 5×10^{-9} - a remarkable sensitivity made possible by the use of a sensitive torsion balance[2]. In 1955, when Lee and Yang looked at the possibility of existence of new long range forces which could be generated by conserved quantities like the baryon number, they used Eötvös' null results to constrain the strength of such a long range force[3].

The realization that the existence of the infinite range (zero mass) electromagnetic field is closely linked to the conservation of electric charge is the basic idea which prompted several physicists to suggest, at various times, the existence of other long range fields coupled to conserved charges like the baryon number. Lee and Yang discussed the possibility of a baryon gauge transformation and associated gauge field. They argued that, since baryon number is not strictly proportional to mass (the fractional variations, induced by nuclear binding, are of the order of 10^{-3}), this new force would be composition dependent. Since it was known from the Eötvös experiments that composition dependent variation in the accelerations of massive bodies towards the Earth are smaller than 10^{-8}, they were able to constrain the strength of the new force to less than 10^{-5} of the gravitational coupling strength between baryons.

The other theoretical implications of a null result in Eötvös type experiments was elaborated by Dicke[4] in several contexts and also by other authors later. Other than providing solid foundation to the metric theories of gravity this null results also has implications on spatial variations (or variations as a function of a scalar field) of fundamental constants (like α, the fine structure constant) and fundamental ratios like the proton-neutron mass ratio. Using Dike's limit for the differential acceleration between gold ($E_b/Mc^2 = 0.004$) and aluminium ($E_b/Mc^2 = 0.001$), 10^{-11}cm/sec^2, we can obtain an upper limit of 10^{-30}/cm for the quantity $\frac{1}{\alpha}\frac{\partial\alpha}{\partial x}$, a very small quantity indeed and a much better limit than what is possible from direct experiments. This kind of a limit on the spatial variation of the fundamental constants is important also from the point of view of considering the universe as spatially homogeneous and isotropic.

Renewed interest in the possible existence of new fundamental interactions developed when various extensions of the standard model were seen to result in the prediction of new long range forces[5]. Supersymmetric solution to the hierarchy puzzle also brings in new fundamental particles and in some models there are new forces mediated by spin-1 particles. There are models which attempt to solve the problem of the cosmological constant by invoking a new field which helps dynamically tune the cosmological constant to near zero value and this model predicts a composition dependent force. This force couples to a linear combination of mass and smaller contributions of baryon number and lepton number and hence violates UFF. There are also predictions of new forces from the axion solution to the strong CP puzzle (spin dependent force), Kaluza-Klein type quantum gravity models and supergravity models. Supergravity models pioneered by Scherk predicts a UFF violating force coupling approximately to B-0.17L with a strength of 0.03% of gravity. Another important charge which could be a source of a new field is B-2L (or the nuclear isospin). A new force mediated by the dilaton arising from broken scale invariance was extensively discussed by Fujii[6] and was an important factor in generating renewed interest in new macroscopic range forces.

Gibbons and Whiting[7] examined satellite and geophysical data and made the point that there was very little scope for a theory which allowed more than 1% deviation from Newtonian gravity at laboratory distances or larger length scales. (See ref. 11 for more details and discussion.)

More recently Damour and Polyakov have found that it is possible for the weak coupling dilaton field from string theories to remain massless through cosmic evolution and there could be observable violations of composition independence of gravity[8] at the level of $10^{-14} - 10^{-23}$. This is an important observation since there are new tests of the equivalence principle planned with sensitivity possibly down to 10^{-17} in a space experiment and a study at the 10^{-14} level might be possible even in a terrestrial experiment. The importance is that this might be the only available experimental probe into string physics at present.

A finite range fifth force coupling to charge Q is described by the Yukawa potential and along with the gravitational potential it is

$$V_{tot} = \frac{-Gm_1 m_2}{r}\left[1 - \alpha\, q_1\, q_2 \exp\left(-r/\lambda\right)\right]$$

between two bodies of mass m_1 and m_2. In this expression, α is the coupling strength in units of gravity ($\alpha = f^2/4\pi\, Gm_H^2$ where m_H is the mass of the hydrogen atom), λ is the range of the force and q is the charge per unit mass (Q/m), the *specific charge*. We assume that gravity itself is composition independent, a fact tested to an accuracy of 10^{-12} in modern Eötvös type torsion balance experiments by Dicke and Braginsky[9]. The important charges under consideration are the Baryon number (B), Lepton number (L) and linear combinations of these charges like the nuclear isospin ($I \equiv N - Z = B - 2L$ for normal nuclei) since these have considerable theoretical significance.

Considerable interest in experimental study of new forces was inspired by indications in the old Eötvös data and in the data from some mine experiments to measure the value of the Newtonian gravitational constant, of the existence of a new force coupling to baryon number with a strength of about 1% of gravity and a range between 100m and 1km[10].

All the experiments performed so far have failed to see any evidence for the existence of a new force coupling to any of the important charges for ranges above 10cm and strength as low as 10^{-4} to 10^{-5} of gravity[11]. The most sensitive Galilean free fall experiments where two test masses are dropped in vacuum chamber and their differential displacement is monitored have obtained an upper limit of about a meter for the product of the range and strength, i.e., $\alpha\lambda \leq 1m$. For ranges less than 10km or so, torsion balance experiments have better sensitivity.

Torsion balance experiments

The popularity of the torsion pendulum as the basic element in gravitation experiments arises from the great convenience and sensitivity it provides for detecting a composition dependent interaction. It is also used extensively in experiments which look for new forces coupling to mass or equivalently to look for violations of the Newtonian inverse square law of gravitation. The advantages offered by a well designed torsion balance are many-fold: (a) The sensitivity can be extremely large, normally limited by internal friction in the fibre support. (b) Earth's gravitational field acts along the fibre axis and therefore does not exert any torque on the pendulum along the fibre, and this property eliminates a potential noise source, (c) The mass elements which constitute the torsion pendulum can be shaped and symmetrized to minimize couplings of multipole moments to gradients in the gravitational field due to the source masses, (d) The motion of the pendulum is periodic and initial conditions are not important in deciding the systematic noise in the experiment, in contrast to Galilean free fall experiments where fine tuning of initial conditions of the two test masses is necessary, (e) The presence of the force can be made to influence the amplitude, phase or the frequency of the pendulum and such a wide choice provides great amount of measurement convenience and possibility for cross checking, (f) It is possible to modulate the basic signal at the natural frequency (resonance) or at another convenient frequency (modulation can be achieved by moving the source masses or by rotating the torsion pendulum) and this increases the signal to noise ratio considerably. Also the observation time can be very long for a high-Q pendulum, which also helps in increasing the signal to noise ratio.

Conceptually, the torsion pendulum for fifth force experiments consists of a massive body made of two compositionally different materials - a compositional dipole - suspended by means of a torsion fibre (usually quartz or tungsten.) The choice of the two materials forming the compositional dipole is decided by the 'charge' to which the new force is supposed to couple. In the simplest experiments, the angular deflection of the pendulum in the presence of a source mass is measured using an optical level or a capacitance transducer.

In the original proposal for starting gravitation experiments in India, some novel designs for the torsion balance were described[12]. One important concept was that of a gravitational monopole and a compositional dipole. Such a mass element will not couple to gradients in the gravitational field while responding to composition dependent forces. In experiments searching for composition dependent forces, this is what is ideally needed since the major source of systematic noise in such experiments is the coupling of the multipole moments of the torsion balance to gradients in the gravitational field of the source masses. A ring made of two semi rings of two different materials satisfy this criteria if the balance and the source masses are constrained to be in a plane. This design was employed in our experiments, with copper and lead as the compositionally different materials. These experiments have been performed in three phases, with the first phase starting in 1987 and the last phase completing in 1992 (see ref.13 for details on design and implementation). These were resonance experiments in which the source masses made of lead and brass are moved around the ring shaped mass element of the torsion balance *at the natural period of the pendulum*, and the changes in the amplitude of the torsional oscillations are monitored to see the influence of a composition dependent interaction.

The result from the phase I experiment was important in ruling out the existence of a new force coupling to isospin[14]. The first few experiments from various groups had conflicting results with a possibility of reconciliation if there was a force coupling approximately to the charge B-2L. But this possibility was found untenable since our experiment and simultaneously another experiment from the Eöt-Wash group in Seattle[15] did not see any evidence for such a force down to a sensitivity of 10^{-3} for α. In phase II, the overall sensitivity of the experiment increased tenfold and provided the most significant constraints on the strength of a new force coupling to isospin[16]. Further improvement in sensitivity by a factor of five was achieved in phase III, by eliminating most of the gravitational gradients[17]. The acceleration sensitivity of the torsion pendulum employed is among the highest in the world and it is about 10^{-12} cm/sec^{-2}.

A high sensitivity optical lever with a resolution of 3×10^{-9} $rad/\sqrt{Hz}$ measured the angular position of the pendulum and the average rate of change of amplitude during the resonant drives translates into values for various coupling constants.

The 2σ upper limit on the strength of the Isospin coupling is given by

$$|\alpha_I| \lesssim 5.9 \times 10^{-5}$$

This limit is applicable beyond a range of 2 m or so and below this range the numbers are to be scaled appropriately by taking into account of the Yukawa character of the potential, with a source to detector distance of 1.2 meters. At 1m, the limits are higher by a factor of 1.5 and for $\lambda = 20$ cm, the upper limit on α is less than 0.4% of gravity.

We can also derive upper limits on the strength of a new force coupling to other important charges like L and $B-L$. These limits are approximately equal and are given by

$$|\alpha_{B-L}| \approx |\alpha_L| \leq 2.32 \times 10^{-4}$$

The combined results from several fifth force experiments indicate that there is no new force coupling to a linear combination of B and L (mainly to B, L, $B-L$ and $B-2L$) or to mass with strength more than 10^{-5} to 10^{-4} of gravity and with range more than a few meters. This rules out conclusively the hypothesis suggested by Fischbach et al, of a fifth force with coupling strength around 1% of gravity. The results from fifth force experiments constrain considerably the theories we discussed earlier. In supergravity theories or in Kaluza - Klein type theories, the coupling strengths of new intermediate range vector interactions should be constrained below 10^{-4} of the gravitational coupling strength. This means that the 'natural' coupling strengths (~ gravitational coupling strength) of the vector or scalar partners of the gravitation in the unbroken theory, need to be scaled down by a factor of at least 10^3 to 10^4 by some mechanism when the low energy behaviour is considered if the mass of the corresponding exchange boson is less than 10^{-7} eV.

All the major groups are now upgrading their detectors and facilities to conduct more sensitive experiments. The Eöt-Wash group has operated a rotatable torsion balance on the turntable with a 3-ton Uranium source mass in close proximity to constrain α_I below 5×10^{-6}. Other than proposed tests of the equivalence principle there are several planned experiments to explore the validity of the Newtonian inverse square law of gravitation, specifically at short distances. This is the direct probe to the existence of new forces coupling to mass. The most sensitive of these experiments employs a three axis superconducting gravity gradiometer designed by H. J. Paik[18]. This set up along with a pendulum source mass weighing 1500 kg was used to obtain a very stringent limit of $\alpha = 0.9 \pm 4.6 \times 10^{-4}$ at $\lambda = 1.5\,m$. Boynton's group at Seattle has designed a torsion pendulum mass element which is sensitive only to horizontal component of the gradient of the Laplacian and this couples to the third order ($n = 3,\ l = 1$) mass moment of the pendulum. The strategy is based on the observation that for a torsion pendulum experiment the lowest order detectable signature which is distinctive of non-Newtonian character of a new interaction is proportional to the gradient of the Laplacian of the interaction potential. The projected range of α is around 10^{-4} for ranges above 10 cm. Experiment which employ the Superconducting gradiometer basically measures the radial derivatives of the Laplacian whereas in Boynton's experiment azimuthal derivatives are measured. Newman's group at Irvine is developing high-Q torsion balances which are cooled to liquid helium temperatures and the goal is to reach a sensitivity of 10^{-14} for a test of the WEP[18].

New techniques and possibilities

So far we have been considering only systems with negligible gravitational binding energy. A statement of the Weak Equivalence Principle when test bodies with considerable gravitational binding energy are involved is called the Gravitational WEP (GWEP) and the corresponding Einstein generalization is called the Strong Equivalence Principle (SEP). The limit on the violation of GWEP is from the lunar laser ranging experiment. The laser ranging experiment monitors the orbital distance of the moon relative to the Earth, with an rms accuracy of several centimeters (the most recent results have an rms error of only about 15 mm). Since the gravitational binding energy of the earth and the moon are considerably different, it is possible to get useful constraints on the violation of GWEP from these experiments. The important point to note here is that to derive any conclusion on the nature of free fall of the gravitational binding energy, one has to assume that the WEP is valid – that the Equivalence principle is valid for *all forms of energy momentum* except possibly the gravitational binding energy. To make this assumption, the laboratory experiments, where the gravitational binding energy is completely negligible, have to establish the WEP to the required accuracy. Right now the sensitivity of the best laboratory experiment is comparable to the lunar laser ranging result and the latter is steadily improving. This means that there is an urgent need to take up new laboratory experiments which can keep pace with the possibly surpass the sensitivity achieved in the laser ranging experiment. The required improvement in sensitivity is a factor of 10 to 100, in the next three to ten years.

Experiments which test the WEP or the GWEP have macroscopic bodies as test particles and the WEP for the elementary particles which constitute the macroscopic body is inferred. This is satisfactory for particles but need not be entirely so in the case of antiparticles. Schiff had presented the argument that since there are clouds of virtual particle-antiparticle pairs in the vacuum polarized space surrounding particles in macroscopic bodies experiments which use macroscopic test bodies to test EP have good sensitivity to address the question of the equivalence of inertial and gravitational mass for antiparticles as well[19]. Experiments have been performed on photons, electrons, neutrons and atoms to check the validity of EP directly for elementary particles. There are ongoing efforts to test the EP directly for antimatter such as antiprotons and antihydrogen.

In this context it is important to mention an emerging area of great potential and possibilities, namely that of interferometry with quantum systems, especially low energy atoms. Neutron interferometry has already proved useful in gravitation experiments to some limited extent. In fact measured phase shifts from neutron interferometry was used to put constraints on a long range fifth force. In comparison to neutron interferometry, atom interferometry has several additional advantages in the context of gravitation experiments. The most important of these are 1) possibility to do experiments with different atoms thereby enabling the study of composition dependent effects 2) ability to trap and cool atoms to ultra low temperatures which increases sensitivity to external perturbations like the gravitational field (the atomic fountain set up can now measure $\Delta g/g \sim 10^{-8}$ which is comparable to the sensitivity of other types of accelerometers). 3) Their large mass and equivalent small wavelengths lead to large amplification of phase perturbing effects 4) ability to manipulate atomic beams using differences in their internal states which allows beam splitting and bending.

Other than exploiting atomic beam techniques and interferometry for the study of equivalence principle there also the possibility of searching for short range forces using these techniques. These experiments, for example, could measure the transmission of specific atomic beams through microgaps and it is conceivable that short range potentials affect the propagation in a measurable way. Modern experiments have provided very stringent constraints on the strength of new feeble forces for ranges beyond a centimeter or so. But, it is important to note that there are no serious constraints on the coupling strengths of new forces for ranges below 1 mm and in the micrometer region new forces with strength a billion times that of gravity cannot be ruled out! It is a challenging area for the experimental physicists, to conceive and execute precision experiments to probe this region. Our group is planning a new series of experiments with neutral atomic beams propagated between parallel plates with very small gap. Normally the transmission of the beam will be affected dominantly by short range forces like the Vander Waals force and the Casimir force[20], along with any new short range force. Modulation of the gap between the mirrors would enable a phase sensitive detection scheme of the forces acting on the propagating beam. One might be able to extract some information on the strength of short range composition dependent forces by choosing suitable elements for the atomic beams when all the known contributions are subtracted out.

Though torsion balances remain the best transducers for measuring differential accelerations in gravitation experiments, there are several problems in increasing their intrinsic sensitivity within current design strategies. One of the major problems to be tackled is the difficulty in realizing a pendulum of large mass, and also of very long period and hence low thermal noise and better sensitivity. The requirements of high load-bearing capacity *and* of a low torsion constant of the suspension fibre are contradictory.

We have been experimenting with a torsion balance set up in which gravitational torques are configured to act against the restoring torque from the suspension fibre, thereby generating lesser restoring force than what would be intrinsic to the fibre with a specified weight bearing capability. The negative restoring set up makes use of the horizontal gradients in the horizontal gravitation field generated by fixed masses juxtaposed with the mass elements of a Cavendish torsion balance, with all the masses lying more or less in a single plane[12,17]. The equilibrium position is determined by the combined torques from the fibre and the fixed masses and if there is any deflection from this position the fibre would act to restore it to equilibrium whereas the gravitational torques, being derived from attractive force, act in the direction to increase the deflection providing the negative restoring force.

With the torsion constant of the fibre thus effectively reduced, it is possible to attain *arbitrarily large periods* provided the natural period of the pendulum without the negative restoring scheme is large compared to the quantity $1/\sqrt{G\rho}$, where ρ is the effective density of the distributed masses in the system. Experiments with a 300 g dumb-bell balance with a suspension which was rather stiff (to avoid the difficulties associated with enormous sensitivity to very small torques, since the first measurements were done in an enclosure which was not evacuated) gave a 5% increase in period of a 100 sec pendulum. To achieve bare periods larger than $1/\sqrt{G\rho}$ without sacrificing the advantages of using a heavy mass element for the torsion balance, we are experimenting with tungsten ribbons as suspension elements. Compared to fibres with circular cross section ribbons with rectangular cross section can have significantly lower torsion constant for the same cross sectional area. The ratio of the torsion constants for the two cross sections of same area scales approximately as $1:\frac{2t}{w}$ where t is the thickness and w

is the width of the ribbon. The 105 μ diameter fibre when replaced by a ribbon which is 30 μ × 300 μ increases the period by about a factor of two with a resulting improvement in sensitivity by a factor of 4, slightly less than expected, but very significant for the future experiments. This torsion balance has also shown improved drift characteristics enabling continuous monitoring for longer durations. Coupled with an improved optical lever with a sensitivity of 3×10^{-10} rad. for angle measurements, the same fifth force search could now be done with a sensitivity approaching 5 parts in a million for α_I and 5×10^{-4} for α_B. This requires suppression of any residual systematic effects that would be visible at the improved sensitivity.

With increased sensitivity the balance is susceptible to various systematic effects and the advantage of higher sensitivity is useful only if these systematic effects are suppressed. Passive shielding of various disturbances is one possibility. At present we are assembling a set up in which the torsion balance is housed in two cascaded vacuum chambers, a ultra high vacuum chamber inside a high vacuum chamber, with additional airtight sealing inside the experimental pit to provide immunity from variations in atmospheric pressure and temperature. This is expected to provide enough passive shielding to perform a test of the WEP at the level of 10^{-13}. Another scheme which is proposed is to employ a differential design. Normally torsion balance experiments are performed with a single torsion balance. If there are uncontrollable systematic effects larger than the signal averaging over a long observation period is not going to reduce the noise. Minimally two pendulums with almost the same characteristics are required, configured with their compositional axes 180° out of phase. Then the two composition dependent signals will be out of phase whereas the systematic effects from temperature variations etc. are expected to be more or less similar enabling a noise subtraction throughout the time of observation. This idea is being pursued to the design level at present.

The outcome of all these preparations is expected to be a very sensitive test of the equivalence principle at the level of 10^{-14}. This is an improvement of a factor of more than hundred on the previous experiments. There is only one experiment more ambitious than this planned to significantly improve the test of the equivalence principle, among other scientific goals. This is the Satellite Test of the Equivalence Principle (STEP) jointly proposed[18] to NASA and ESA, by a team of physicists led by C.W.F. Everitt. This space experiment on board a drag free satellite is a modern version of the Galileo experiment. The payload consists of six differential accelerometers for testing the WEP, two accelerometers for measuring the gravitational constant and for testing short range inverse square law and one accelerometer to search for a new force with coupling between normal and spin polarized matter (inspired by axion hypothesis.) The projected sensitivity for the equivalence principle test is about 10^{-17} and for the measurement of G and for the test of inverse square law the sensitivity is expected to be 10^{-6}. The satellite can be launched only around 2005 and till then terrestrial experiments are the only ones which can provide improved constraints on composition dependence in gravity.

The initial ambitions in starting gravitation experiments in India included the detection of galactic dark matter, possibly in the form of massive neutrinos, using a suitably prepared high sensitivity torsion balance. The sensitivity needed for such a task is a million times more than what is presently achieved, but the ideas involved are still very attractive. At present the most popular particle candidates for dark matter are low mass neutrinos and heavy supersymmetric particles. There are several experiments in progress to directly detect the heavy WIMPs (Weakly Interacting Massive Particles) using calorimetric and ionization detection techniques employing cryogenic single crystal detectors[21]. There are useful constraints though the required signal to noise ratio for a direct detection is estimated to be a factor of 100 to 1000 higher than what is presently available.

There have been alternatives to the dark matter hypothesis, and one such important class of suggestions is called Modified Newtonian Dynamics (MOND)[22]. The idea is that either the law of inertia or the law of Newtonian gravity is modified below a critical acceleration a_0, typically of the order of 10^{-8} cm/sec^2. The suggested modification is non linear; $F = m(aa_0)^{1/2}$ where 'a' is the normal Newtonian acceleration. This, for small accelerations typical of galactic halo regions gives flat rotation curves. The phenomenelogical ideas have been subsequently elevated to the status of a field theory of Milgrom and Bekenstein[23]. It is tempting to think of a direct test of this theory[24] since the maximum intrinsic accelerations during the free torsional oscillations of our pendulum is one order less than a_0! Also the pendulum can measure accelerations about three to four orders smaller than a_0. But a conceptual difficulty arises because the MOND does not obey the Strong Equivalence Principle. Since the torsion pendulum is already in the field of local masses, sun and the galaxy and so on which are all larger than a_0, it is argued that MOND can never be tested in the laboratory, since the addition law for accelerations is nonlinear in the theory spoiling the possibility of a null experiment. It would be very significant if some way of getting around this difficult could be envisaged.

In conclusion I would like to stress the point that the non-accelerator experiments described in this talk are among the small number of experimental probes available to study physics beyond the standard model and that torsion balance experiments have a special role to play in this class of experiments.

Acknowledgements

I thank all members of the gravitation experiments group – Prof. R. Cowsik, Prof. S.N. Tandon, Dr. N. Krishnan, U.D. Vaishnav, S.M. Periera, P.G. Rodrigues, A.B. Patil, D.B. Mane, A. Vaidyanathan, P.K.S. Murthy, C. Rajanna and K.C. Nallaraju. Most of the experiments from our group were conceived and proposed by Prof. R. Cowsik. Support provided by the staff of the Seismic Array Station (BARC), Raman Research Institute and Indian Institute of Astrophysics at Gauribidanur is also gratefully acknowledged.

References

1. C.M. Will, *Theory and Experiment in Gravitational Physics*, Cambridge University Press (1981).
2. R. von Eötvös, D. Pekár and E. Fekete, *Ann. Phys. (Leipzig)*, **68**, 11, (1922).
3. T.D. Lee and C.N. Yang, *Phys. Rev.*, **98**, 1501, (1955).
4. R.H. Dicke, Theoretical Significance of Experimental Relativity, Gordon and Breach, New York, (1964).
5. J. Scherk, *Phys. Lett.*, **88**, 265, (1979); J.E. Moody and F. Wilczek, *Phys. Rev.*, **D30**, 130, (1984); T. Goldman, R.J. Hughes and M.M. Nieto, *Phys. Lett.*, **B 171**, 217, (1986); D. Chang, R.N. Mohapatra and S. Nussinov, *Phys. Rev. Lett.*, **55**, 2835, (1985).
6. Y. Fujii, *Nature (Phys. Sci.)*, **234**, 5, (1971); *AAPPS Bulletin*, **4**, No. 2, (June 1994), 19.
7. G.W. Gibbons and B.F. Whiting, *Nature*, **291**, 636, (1981).
8. T. Damour and A.M. Polyakov, *Nucl. Phys.*, **B 423**, 532, (1944).
9. P.G. Roll, R. Krotkov, and R.H. Dicke, *Ann. Phys. (NY)*, **26**, 442, (1964); V.B. Braginsky and V.I. Panov, *Sov. Phys. JETP*, **34**, 463, (1972).
10 E. Fischbach, D. Sudarsky, A. Szafer, C. Talmadge and S.H. Aronson, *Phys. Rev. Lett.*, **56**, 3, (1986); Ann. Phys. **182**, 1, (1988).
11. For a brief review see C.S. Unnikrishnan, Pramana - *J. Phys.* (supplement issue - Proceedings of the X DAE symposium on High Energy Physics), **41**, 395, (1993).
12. R. Cowsik, *Indian J Physics*, **B 55**, 497, (1981); *Challenges in experimental gravitation and cosmology – DST proposal* (1982).

13. R. Cowsik, N. Krishnan, P. Saraswat, S.N. Tandon, C.S. Unnikrishnan, U.D. Vaishnav, C. Viswanadham and G.P. Puthran, *Ind. J. Pure. Appl. Phys.*, **27**, 691, (1989); N. Krishnan, *Ph.D Thesis*, University of Bombay / TIFR (1989), unpublished; C.S. Unnikrishnan, *Ph.D Thesis*, University of Bombay / TIFR (1992), unpublished.
14. R. Cowsik, N. Krishnan, S.N. Tandon, C.S. Unnikrishnan, *Phys. Rev. Lett.*, **61**, 2179, (1988).
15. The Eöt-Wash experiments are disscussed in detail in E.G. Adelberger, B.R. Heckel, C.W. Stubbs, and W.F. Rogers, *Ann. Rev. Nucl. Part. Sci.*, **41**, 269, (1991).
16. R. Cowsik, N. Krishnan, S.N. Tandon, C.S. Unnikrishnan, *Phys. Rev. Lett.*, **64**, 336, (1990).
17. C.S. Unnikrishnan, *Classical and Quantum Gravity*, **11**, A195, (1994).
18. See Proceedings of the International Conference on Experimental Gravitation, Nathiagali, Pakistan (Ed. M. Karim) – special issue of *Classical and Quantum Gravity*, **11**, (1994).
19. L.I. Schiff, *Phys. Rev. Lett.*, **1**, 254, (1958).
20. V. Sandoghdar, C.I. Sukenik, E.A. Hinds and S. Haroche, *Phys. Rev. Lett.*, **68**, 2432, (1992); C.I. Sukenik, M.G. Boshier, D. Cho, V. Sandoghdar and E.A. Hinds, *Phys. Rev. Lett.*, **70**, 560, (1993).
21. See R. Cowsik, *Current Science*, **61**, 759, (1991) for a general review; Details of detection techniques and cross sections were discussed by J.R. Primack, D. Seckel and B. Sadoulet, *Ann. Rev. Nucl. Part. Sci.*, **38**, 751, (1988).
22. Milgrom. M, *Ap. J.*., **270**, 365, (1983).
23. Bekenstein. J.D. and Milgrom. M, *Ap. J.*, **286**, 7, (1984).
24. C.S. Unnikrishnan, paper presented at the annual meeting of the Astronomical Society of India, IUCAA, Pune, (1994), to be published.

A Model of the Void in an Expanding Universe

D.R. Mandal* **and S. Banerji****

*Department of Physics, Basirhat College, Basirhat, North 24 - Parganas, W. Bengal, India.

**Department of Physics, The University of Burdwan, Burdwan 713 104, W. Bengal, India.

Abstract

We consider here a model of the spherical void containing low density conducting fluid surrounded by a thick spherical shell and radiation embedded in a Robertson-Walker (RW) universe with flat space sections. The void has a metric which is the special case or a solution given by Maiti (1982) surrounded by Vaidya metric. We also assume the R-W universe to be filled with a perfect fluid with a linear equation of state. The matching conditions indicate that the void goes on contracting as the universe expands until it collapses to a point. However, if the pressure in the external universe vanishes, the void remains static.

1. Introduction

Astronomical observations in the last decade have indicated the existence of regions in the universe which appear to be empty, called voids. Later evidence indicated that the voids are deficient in luminous matter but are not completely empty [see Bonnor and Chamorro.A. 1990 and the references therein]. Voids of sizes upto 60 h^{-1} Mpc (H_0 - 100 h Kms^{-1} Mpc^{-1}) have been observed. Many theoretical investigations into the origin and evolution of such voids have been published. There have been four principal lines of investigation : small perturbations of homogeneous cosmologies, use of the Einstein-Strauss Vacuole, treatment of the boundary of the void by thin wall approximation and matching of the exact solutions of the Einstein's equations including Tolman spacetimes with a homogeneous model of the expanding universe on the outside. The present paper is a representative of the last named approach.

Here we shall give a model of the void consisting of a core surrounded by a thick spherical shell of pure radiation. The metric inside the core (Region I) is a special case of that given by Maiti (1982). The outer shell of radiation has the metric given by Vaidya (1953) (Region II). This combination has been matched with the Robertson-Walker (RW) metric with zero spatial curvature. For simplicity we assume the RW universe to be filled with a perfect fluid with a linear relation between pressure p and the density ρ so that the scale factor has the form $\sim t^n$. In the early radiation dominated universe $n = 1/2$ while $n = 2/3$ for pressureless dust, which approximates the present condition.

In section 2, we shall given an account of the metrics in the three different regions and the relevant field equations. In section 3 we shall discuss the boundary conditions and their implications. In section 4 we shall discuss the main conclusions.

2. A model of the void

The core of the Void called Region I has a metric of the form given by Maiti (1982):

$$ds_1^2 = \left[1 + \frac{a}{1+\xi r_1^2}\right]^2 dt_1^2 - \frac{R^2(t_1)}{(1+\xi r_1^2)^2}\,(dr_1^2 + r_1^2 d\theta^2 + r_1^2 \sin^2\theta\, d\varphi^2) \quad (2.1)$$

where a and ξ are both constants. The energy momentum tensor is that of a fluid with heat flux expressed in the standard form as

$$T_\mu^\nu = (\rho + p)\, u_\mu u^\nu - p\,\delta_\mu^\nu - q_\mu u^\nu - u_\mu q^\nu \quad (2.2)$$

where q^μ represents the heat flux vector which is orthogonal to the velocity vector u^μ. In the present spherically symmetric case the radial component q^1 is nonvanishing. In comoving coordinates Einstein's field equations for the metric (2.1) are :

$$8\pi p = -\frac{4\xi}{R^2} - \frac{4a\xi}{R^2}\,\frac{1-\xi r_1^2}{1+\xi r_1^2}\left(1 + \frac{a}{1+\xi r_1^2}\right)^{-1} - \left[2\frac{\ddot{R}}{R} + \left(\frac{\dot{R}}{R}\right)^2\right]\left(1 + \frac{a}{1+\xi r_1^2}\right)^{-2} \quad (2.3)$$

$$8\pi\rho = \frac{12\xi}{R^2} + 3\left(\frac{\dot{R}}{R}\right)^2\left(1 + \frac{a}{1+\xi r_1^2}\right)^{-2} \quad (2.4)$$

$$8\pi q^1 = -\frac{4a\xi r_1 \dot{R}}{R^3}\left(1 + \frac{a}{1+\xi r_1^2}\right)^{-2} \quad (2.5)$$

By taking suitable values of ξ, a and R we can make p and ρ small.

In Region II the metric is taken in the form of Vaidya (1953).

$$ds_2^2 + \left(1 - \frac{2m(v)}{r_2}\right) dv^2 + 2dv\, dr_2 - r_2^2\, (d\theta^2 + \sin\theta\, d\Phi^2) \tag{2.6}$$

In Region III we take the Robertson-Walker metric with flat space sections filled with a perfect fluid with $p = \gamma\rho, 0 \le \gamma \le (1/3)$. The corresponding solution involves a power in t.

$$ds_3^2 = dt_3^2 - t_3^{2n}\, (dr_3^2 + r_3^2 d\theta^2 + r_3^2 \sin^2\theta\, d\varphi^2) \tag{2.7}$$

For $\gamma = 0$ (dust), $n = 2/3$
For $\gamma = 1/3$ (pure radiation), $n = 1/2$

3. Boundary Conditions

The matching conditions on the boundary between Regions I and II obtained by equating the first and second fundamental forms on the two sides are given below.

$$A^2 \dot{t}_1^2 - B^2 \dot{r}_1^2 = 1 \tag{3.1}$$

$$r_2 = Br_1 \tag{3.2}$$

$$1 - \frac{2mv}{r_2}\quad \dot{v}^2 + 2\dot{r}_2 \dot{v} = 1 \tag{3.3}$$

$$r_1^2 A\dot{t}_1 \frac{\partial B}{\partial r_1} + r_1 AB\dot{t}_1 + \frac{B^2 r_1^2 \dot{r}_1}{A} \frac{\partial B}{\partial t_1} = -\left(1 - \frac{2m}{r_2}\right) \dot{v} r_2 - r_2 \dot{r}_2 \tag{3.4}$$

$$\ddot{t}_1 AB\dot{r}_1 - \ddot{r}_1 AB\dot{t}_1 - A\dot{t}_1 \dot{r}_1^2 \frac{\partial B}{\partial r_1} - \frac{A^2}{B} \dot{t}_1^3 \frac{\partial A}{\partial r_1} + 2Br_1^2 \dot{t}_1 \frac{\partial A}{\partial r_1}$$
$$- 2Ar_1 \dot{t}_1^2 \frac{\partial B}{\partial t_1} + \frac{B^2 \dot{r}_1^3}{A} \frac{\partial B}{\partial t_1} = \frac{m\dot{v}}{r_2^2} - \frac{\ddot{v}}{\dot{v}} \tag{3.5}$$

where $A = 1 + \dfrac{a}{1 + \xi r_1^2}$, $\quad B = \dfrac{R(t_1)}{1 + \xi r_1^2}$

A dot denotes derivative with respect to τ, the time corresponding to the comoving coordinates in the metric intrinsic to the boundary [Santos 1985].

The matching of the Vaidya metric of Region II with that of RW of Region III has been done by a number of workers [Fayos et al 1992 and Aguirregabiria et al 1991].

The conditions are:

$$r_2 = u t_3^n \tag{3.6}$$

$$\dot{t}_3^2 - t_3^{2n}\, \dot{u}^2 = 1 \tag{3.7}$$

$$\left(1 - \frac{2m}{r_2}\right) r_2 \dot{v} + r_2 \dot{r}_2 + u t_3^n \dot{t}_3 + m u^2 \dot{u} t_3^{3n-1} = 0 \tag{3.8}$$

$$\frac{m\dot{v}}{r_2^2} - \frac{\ddot{v}}{\dot{v}} + \ddot{u} \dot{t}_3\ t_3^n + 2n\, \dot{u}\, \dot{t}_3^2\, t_3^{n-1} - m\dot{u}^3\, t_3^{3n-1} - u t_n^n \ddot{t}_3 = 0 \tag{3.9}$$

where $r_3 = u(t_3)$ is the equation of the boundary. We have the following solutions of equations (3.6) - (3.9).

$$u = \frac{3n-2}{3n(1-n)} t_3^{1-n} + \alpha_0, \qquad \alpha_0 > 0 \tag{3.10}$$

$$r_2 = u t_3^n = \frac{3n-2}{3n(1-n)} t_3 + \alpha_0 t_3^n \tag{3.11}$$

$$\dot{t}_3 = \frac{3n}{2(3n-1)^{\frac{1}{2}}} \qquad \text{(taking only the positive sign)}$$

$$m = \frac{n^2}{2} u^3 t_3^{3n-2} \tag{3.12}$$

$$v = \int \frac{\dfrac{2(3n-1)}{9n(1-n)} + \dfrac{2(3n-1)}{3} \alpha_0 t_3^{n-1}}{n^2 \alpha_0^2 t_3^{2(n-1)} + \dfrac{2(3n-2)\, n\, \alpha_0}{3(1-n)} + \dfrac{6n-5}{9(1-n)^2}}\, dt_3 \tag{3.13}$$

It is evident from (3.13) and (3.11) that

$$\frac{dm}{dt_3} = \frac{n(3n-2)}{2} u^2 t_3^{3(n-1)} \frac{t_3^{(1-n)}}{(1-n)} + nu) \tag{3.14}$$

This is negative for $\frac{1}{2} \leq n \leq \frac{2}{3}$ and zero for $n = 2/3$. Now obviously u decreases with increase of t_3. The Void therefore contracts until $u \to 0$. If, however n changes with time and reaches the value $n = 2/3$ appropriate to the present epoch

then $u \to \alpha_0$ the void ceases to contract. In this case Vaidya metric reduces to that of Schwarzschild with constant m and the time coordinate t_2 is given by

$$v = t_2 - \int \frac{dr_2}{1 - \frac{2m}{r_2}} \tag{3.15}$$

4. Conclusions

In this model of the spherical void, the radius of the void formed in the early universe goes on contracting as the universe expands. If, however, the void persists till the present epoch of zero pressure, then it becomes static with a constant radius.

Reference

1. Aguirregabira, J.M. , Ibanez, J. Di. Prisco, A and Herrera, L (1991), Asto-phys. J. **376**, 662.
2. Bonnor, W.B. and Chamorro, A. (1990), Astrophys. J. **361**, 21.
3. Fayos, F. Jaen, X., Llanta, E. and Senovilla, J.M.M. (1992), Phys. Rev. **45D**, 2732.
4. Maiti, S.R. (1982), Phys. Rev. **D25**, 2518.
5. Santus, N.O. (1985), M.N.R.A.S. **216**, 403.
6. Vaidya, P.C. (1953), Nature, **171**, 260.

Understanding Nonlinear Gravitational Clustering in the Expanding Universe

T. Padmanabhan
Department of Astronomy and Astrophysics
The Pennsylvania State University
525 Davey Laboratory, University Park
PA 16802-6305, USA

Permanent address: Inter-University Centre for Astronomy and Astrophysics, Post Bag 4, Ganeshkhind, Pune - 411 007, India. Email: paddy@iucaa.ernet.in

Abstract

The gravitational clustering of collisionless particles in an expanding universe poses several interesting theoretical questions. I discuss some aspects of this problem and argue that it may be possible to understand the nonlinear clustering in terms of some simple physical arguments. This analysis leads to results which are broadly in agreement with numerical simulations.

Large-scale structures in the universe (like galaxies, clusters etc.) are believed to have formed through the growth of small initial density perturbations. The driving force behind the formation of such structures in the universe is the gravitational field produced by inhomogeneities. Overdense regions accrete matter at the expense of underdense regions allowing inhomogeneities in the universe to grow. Observations also suggest that the material content of the universe is dominated by dark matter, likely to be made of collisionless elementary particles. In that case, the gravitational force is mainly due to these particles and, to first approximation, we can ignore the complications arising from baryonic physics. The evolution of perturbations is then governed purely by the gravitational force.

When these density perturbations are small, it is possible to study their evolution using linear theory. But once the density contrast becomes comparable to unity, linear perturbation theory breaks down and one must invoke either analytic approximations or direct N-body simulations to study the growth of perturbations. While these simulations are of some value in making concrete predictions for specific models, they do not provide clear physical insight into the process of non-linear gravitational dynamics. To obtain such an insight into this complex problem, it is important to ask whether any simple pattern exists in the gravitational clustering of collisionless particles in an expanding universe. Of course, one requires the numerical simulations to help the search for such patterns and to test the validity of any specific conjecture.

Some elementary conclusions regarding the clustering can be obtained from the direct examination of the relevant equations. We are interested in the scales with $L \ll H^{-1}$ where $H = (\dot{a}/a)$ is the instantaneous Hubble 'constant'. In such a case, we can describe dark matter as a system of particles interacting via Newtonian gravity. The motion of particles will be described by the equations

$$\ddot{x}_i + \frac{2\dot{a}}{a}\dot{x}_i = -\frac{1}{a^2}\nabla\phi; \ \nabla^2\phi = 4\pi G\rho_b a^2\delta, \tag{1}$$

where $x_i(t)$ is the comoving position of the i-th particle, $a(t)$ is the expansion factor, $\rho_b(t)$ is the background density in a matter dominated universe and ϕ is the gravitational potential due to the perturbed density $\delta\rho \equiv \rho - \rho_b \equiv \rho_b\delta$. All spatial derivatives are with respect to the comoving coordinates. In the fluid limit, the same system is described by the equations

$$\dot{\delta} + \nabla\cdot[(1+\delta)u] = 0; \quad \dot{u} + (u\cdot\nabla)u + \frac{2\dot{a}}{a}u = -\frac{1}{a^2}\nabla\phi, \tag{2}$$

where $u \equiv (dx/dt)$. For a universe with density parameter $\Omega = 1$, we can introduce a logarithmic time coordinate $\tau \equiv \ln(t/T)$ [where T is an arbitrary positive constant] and a rescaled gravitational potential $\psi \equiv (4/9)H_0^{-2}(a\phi)$, where H_0 is the current value of Hubble constant. Then equation (1) reduces to

$$\frac{d^2x_i}{d\tau^2} + \frac{1}{3}\frac{dx_i}{d\tau} = -\nabla\psi; \ \nabla^2\psi = \delta, \tag{3}$$

while (2) becomes

$$\frac{\partial\delta}{\partial\tau} + \nabla\cdot[(1+\delta)p] = 0; \ \frac{\partial p}{\partial\tau} + (p\cdot\nabla)p + \frac{1}{3}p = -\nabla\psi, \tag{4}$$

where p is the velocity field corresponding to τ coordinate. All reference to the background spacetime has completely disappeared in these equations and these equations have no intrinsic scale. Any preferred scale in the solution can only come from the initial conditions.

If we now further assume that the initial power spectrum of perturbation is a scale invariant powerlaw with $P(k) = Ak^n$, then the dynamical evolution of such a system cannot exhibit any preferred scale and must possess some universal characteristics. Requirements of convergence, $(n > -3)$, and the generation of a k^4 tail due to discreteness effects restrict the range of n to $-3 < n < 4$. We shall restrict ourselves to this range[1]. [It should be noted that if the equations (2) are studied perturbatively, (spurious) divergences due to long wavelength modes can arise for $-3 < n < -1$. The long wavelength contribution to bulk velocity, for example, has a divergent contribution for $n = -2$ when handled perturbatively. The nonperturbative argument given in the paragraph above clearly shows that these divergences are spurious.]

One immediate consequence of the above analysis is that a simple scaling relation *must* exist for physically relevant variables characterising the dark matter. Since there is no *apriori* scale in the problem, the only natural length at any time is $x_{NL}(a) \propto a^{2/(n+3)}$, the scale which is going nonlinear at the epoch a. Thus all dimensionless physical parameters must be expressible as universal functions of the variable $q \equiv (x/x_{NL})$. For example, we expect the two-point correlation function $\xi(x, a)$ to have the form $\xi(x, a) = f_n[x/x_{NL}(a)]$. The evolution is self-similar,[2] and ξ is completely specified by a single function $f_n(q)$ which - of course - could depend on the index n.

It is, however, not possible to determine the form of this function [except in the extreme linear end] without further assumptions. What is more, the functional form has a fairly strong dependence on the index n. This means that if the original spectrum is not a pure power law, we cannot obtain any useful insight from the above result. To make further progress, it is necessary to obtain a dynamical equation satisfied by the correlation function and concentrate on variables which have only a weak dependence on n.

An equation governing the evolution of ξ is easy to write down. From the conservation of pairs of particles, one immediately obtains the relation[3]

$$\frac{\partial \xi}{\partial t} + \frac{1}{ax^2}\frac{\partial}{\partial x}[x^2(1+\xi)v] = 0 \tag{5}$$

where $v(t, x)$ denotes the mean relative velocity of pairs at separation x and epoch t. We now define an average correlation function inside a sphere of radius x by

$$\bar{\xi} \equiv \frac{3}{x^3}\int_0^x \xi(t, y)\, y^2\, dy \tag{6}$$

and a dimensionless pair velocity $h(a, x) \equiv -(v/\dot{a}x)$. Using these definitions equation (5) can be written as

$$\left(\frac{\partial}{\partial \ln a} - h\frac{\partial}{\partial \ln x}\right)(1 + \xi) = 3h(1 + \bar{\xi}) \tag{7}$$

This equation can be made even more transperent by introducing the variables

$$A = \ln a, \qquad X = \ln x, \qquad D(X, A) = \ln(1 + \bar{\xi})$$

in terms of which we have

$$\frac{\partial D}{\partial A} - h(A, X)\frac{\partial D}{\partial X} = 3h(A, X) \tag{8}$$

This equation[4] is remarkable in several respects and deserves close scrutiny.

To begin with it shows that the driving force for ξ is the [dimensionless] relative pair velocity $h(A,X)$. If the form of this function is known, one can integrate (8) and obtain ξ. So this equation suggests that one should look at the properties of h closely. We shall get back to this feature in a moment.

This equation also leads to another interesting result: Introducing a variable $F = D + 3X$, (8) can be written as

$$\frac{\partial F}{\partial A} - h(A, X)\frac{\partial F}{\partial X} = 0 \tag{9}$$

The characteristic curves to this equation - on which F is a constant - are determined by $(dX/dA) = -h(X, A)$ which can be integrated if h is known. But note that the characteristics satisfy the condition

$$F = 3X + D = \ln[x^3(1 + \bar{\xi})] = \text{constant} \tag{10}$$

or, equivalently,

$$x^3(1 + \bar{\xi}) = l^3 \tag{11}$$

where l is another length scale. When the evolution is linear at all the relevant scales, $\xi << 1$ and $l \approx x$. As clustering develops ξ increases and x becomes considerably smaller than l. One may say that the behaviour of clustering at some scale x is determined by the original *linear* power spectrum at the scale l through the "flow of information" along the characteristics. With this interpretation, one has a glimpse of how large scale power is transferred to smaller scales. The behaviour of $x^3\xi(a, x)$ for small x will determine the range of large scales that can significantly influence the small scales.

It is an interesting feature of gravitational clustering that large scales affect the small scales while the small scale power does not affect the large scale to the same extent. At any given moment we can clump together small scale virialised objects and replace them by another particle of the same total mass. Dynamical evolution at sufficiently large scales should be insensitive to this replacement since the large scales only feel the net (monopole) mass of the virialised structures. In fact, it is this "renormalisability" of the Newtonian gravity which allows one to perform dark matter numerical simulations wherein each simulation particle represents a huge collection of elementary particles making up the dark matter sea. But the large scales do affect the small scales due to the breaking of long wavelength waves of perturbation. For example, if one starts the evolution with negligible power at small scale, nonlinear clustering will generate it. The relation between l and x allows one to understand this feature. When the correlations are small, $x \approx l$; as correlations develop, x becomes considerably smaller than l showing that during evolution progressively smaller scales are affected by the same l. In this sense, the transfer of power in gravitational clustering is from large to small scales just like in fluid turbulence. It is conceivable that some general pattern - like the Kolmogorov spectrum in turbulence - can exist describing this power transfer.

Finally, equation (8) can be integrated in closed form if the original linear spectrum was a power law. In that case, $\bar{\xi}$ is a function of the scaled variable $q \equiv xa^{-2/(n+3)}$ or - equivalently - D only depends on $Q \equiv \ln q = X - 2A/(n+3)$. It follows from (8) that h can also depend only on Q; hence we can express h entirely in terms of D. Thus, for any power law spectrum, h depends on A and X only through D with $h(A,X) = h_n(D[A,X])$. We have put a subscript n to indicate that possibility that the functional form *could* depend on n. Once it is given that $h = h_n(D)$, we can easily integrate (8) and express[4] the true mean correlation function $\bar{\xi}(a,x)$ in terms of the correlation function in the linear theory $\bar{\xi}_L(a,l)$ as:

$$\bar{\xi}_L(a,l) = \exp\left(\frac{2}{3}\int^{\bar{\xi}(a,x)} \frac{dq}{h_n(q)(1+q)}\right) \tag{12}$$

where $l = x[1+\bar{\xi}(a,x)]^{1/3}$. As to be expected, the linear evolution at l influences the nonlinear physics at x. [Incidentally, equation (8) can also be integrated for a somewhat more general form of $h(A,X)$ which are expressible as $g(A)h_n(D)$. In this case, we only have to change variables from A to B with $dB = g(A)dA$ to obtain a solution. Such a form of h arises in universes with $\Omega < 1$.]

Inverting (12) we can express the true correlation function $\xi(a, x)$ in terms of the correlation function calculated from the linear theory $\xi_L(a, l)$ in the form of a relation:

$$\xi(a, x) = U_n[\xi_L(a, l)] \tag{13}$$

where U_n is a function to be determined (which, of course, can be found if h_n is known and vice versa). Once U_n is known we shall be able to find the results of nonlinear evolution using the results of linear theory. Note that the n-dependence of U_n arises from the n-dependence of h_n. We expect this dependence to be very weak and hence a relation of the kind in (13) is likely to hold universally even for spectra which are not pure power laws.

Let us now try to gain some insight into the form of U_n by simple physical arguments. To begin with, we expect $U_n = 1$ in the linear regime where $\xi \lesssim 1$. In the other extreme limit, for $\xi \gtrsim 200$, we expect the formation of virialised systems based on, say, the spherical top hat model. If we assume that the relative pair velocities of virialised systems exactly balance hubble expansion [in the nonlinear regime] then we must have $h = 1$ for $\xi \gtrsim 200$. To keep the discussion slightly more general, let us assume that h_n is some constant of order unity in the nonlinear regime. In that case, (12) immediately shows that $U_n(\xi_L) \propto (\xi_L)^{3h_n/2}$ in the nonlinear regime. The scaling in the intermediate regime $1 \lesssim \xi \lesssim 200$ is more subtle but can be fixed by the following argument: As clustering proceeds by violent relaxation, the preferred configuration is that of isothermal spheres with a density distribution of x^{-2} corresponding to $n = -1$. This suggests that the scaling between ξ and ξ_L in the intermediate regime should allow linear evolution to remain valid for $n = -1$ spectrum. That is, if $\xi_L(a, x) \propto a^2 x^{-2}$, then $\xi(a, x)$ should also scale as $a^2 x^{-2}$ in the quasilinear regime. This fact - and the relationship between l and x - uniquely fixes the scaling in the quasilinear regime to be $U(\xi_L) \propto (\xi_L)^3$. The proportionality constants can be determined by matching the relations between the three regimes. Taking $h = 1$ in the extreme nonlinear regime, this will give

$$\xi(a, x) = \begin{cases} \xi_L(a, l) & (\text{for } \xi_L < 1, \xi < 1) \\ \xi_L(a, l)^3 & (\text{for } 1 < \xi_L < 5.85, 1 < \xi < 200) \\ 14.14\,\xi_L(a, l)^{3/2} & (\text{for } 5.85 < \xi_L, 200 < \xi) \end{cases} \tag{14}$$

This is a remarkable result! It shows that if the assumption of stable clustering is true, then we can express $\xi(a, x)$ as a universal function of $\xi_L(a, l)$ with no reference to the index n. Hence we should be able to use such a result even for spectra which are not strictly power laws as long as the variation is not too drastic.

Numerical simulations indeed suggest [5,2] that such a relation exists. A more precise form of the fitting function is[6]:

$$\xi(a,x)=\begin{cases}\xi_L(a,l) & (\text{for } \xi_L<1.2, \xi<1.2)\\ 0.7\,\xi_L(a,l)^3 & (\text{for } 1.2<\xi_L<6.5, 1.2<\xi<195)\\ 11.7\,\xi_L(a,l)^{3/2} & (\text{for } 6.5<\xi_L, 195<\xi)\end{cases} \tag{15}$$

The striking similarity between (15) and (14) suggests that the theoretical argument leading to (14) is substantially correct. Given the linear correlation function, this relation allows one to compute the nonlinear correlation function and any other quantity derivable from it. The relation (15) is useful only because the dependence of h on n is fairly weak; in fact, for CDM like models, h is reasonably independent of n and behaves as follows: $h(\xi)\approx(2/3)\xi$ for $\xi \ll 1$, reaches a maximum of about 2 at $\xi\approx 20$ and decreases to a constant value of about unity at large ξ. Since the spectral dependence is weak, (15) holds even for spectra which are not pure power laws as long as the effective power law index does not vary too rapidly.

Incidentally, for power law spectra, we can immediately obtain

$$\frac{\xi(a,x)}{a^2x^{-(n+3)}}\simeq\begin{cases}(.7)^{\frac{1}{n+4}}\,a^{-2\frac{n+1}{n+4}}\,x^{\frac{(n+1)(n+3)}{n+4}} & (\text{for } 1\ll\xi\leq 200)\\ (11.7)^{\frac{2}{n+5}}\,a^{-2\frac{n+2}{n+5}}\,x^{\frac{(n+2)(n+3)}{n+5}} & (\text{for } 200\leq\xi)\end{cases} \tag{16}$$

The first line shows that linear evolution is preserved for $n=-1$ spectra even in the intermediate regime. Equation (16) can, of course, be rewritten in terms of our scale invariant variable $q=xa^{-2/(n+3)}$:

$$\xi(a,x)\simeq\begin{cases}(.7)^{\frac{1}{n+4}}\,q^{-\frac{3(n+3)}{n+4}} & (\text{for } 1\ll\xi\leq 200)\\ (11.7)^{\frac{2}{n+5}}\,q^{-3\frac{(n+3)}{n+5}} & (\text{for } 200\leq\xi)\end{cases} \tag{17}$$

Thus we have succeeded in determining the form of $f_n(q)$ discussed earlier on.

While the relation (15) is valid [and very useful] to fair degree of accuracy, it is easy to see that they cannot be *exact*. In reality, we do expect some n dependence for h - and hence for U - especially in the nonlinear regime. To see this, consider a power spectrum with a sharp maximum at $x\approx L$ and very little small scale power. [This is similar to the HDM spectrum]. The first scales which go nonlinear will correspond to clusters with size $x\approx L$. The evolution of the first nonlinear structures in such a model could be well approximated by collapse of a spherical top hat (STH), since there is very little small scale power. For an STH model, one can easily show[2] that h is a monotonically increasing function of the density contrast $\delta \approx \xi$ and $h \approx \xi^{1/2}$ as $\xi \to \infty$. This suggests that h will increase with ξ for a good range of ξ in the HDM-like models until significant amount of small scale power develops. One can reach the same conclusion from (15) as well: If there is very little small scale power in linear theory, then ξ will have to rise far more steeply with ξ_L (in the nonlinear end) than suggested by (15). The study of such an extreme example suggests that there might exist a weak spectrum dependence in the $h(\xi)$ relation even in more moderate cases. Broadly speaking, we expect h to be lower at a given ξ (in the nonlinear regime) as we add more small scale power.

If we take h to be a constant asymptotically for hierarchical models, then we expect this constant value to decrease with increasing n. To determine this n dependence, we can proceed as follows: One may invoke the hypothesis that the physics in the highly nonlinear regime should have no memory of the initial spectrum and that the highly nonlinear, virialised, system can be described in a manner independent of the original linear power spectrum. When h is a constant, the general solution to (8) is

$$\xi(a, x) = a^{3h} G(a^h x) \tag{18}$$

where G is an arbitrary function of its argument. If we further take the linear spectrum to be a power law with index n, then we must have

$$\xi(a, x) \propto a^{(3-\gamma)h} x^{-\gamma}; \quad \gamma = \frac{3h(n+3)}{2 + h(n+3)} \tag{19}$$

This shows that, apriori, the index in the nonlinear end (γ) "remembers" the index in the linear end (n). How strong this memory is depends on the n-dependence of the combination $h(n+3)$. If we demand that the extreme nonlinear end of the correlation function should be independent of n, then we must have the scaling

$$h = \frac{3c}{n+3} \tag{20}$$

in the nonlinear end with some constant c. The value of $c = 1$ is suggested by the following argument: The case of $h = 1$ corresponds to stable clustering in which the relative pair velocities exactly match the hubble expansion allowing the virialised systems to remain stable. Such a scenario requires careful balance between virialisation and mergers in an $\Omega = 1$ universe. One can expect this balance to exist for pure white noise with $n = 0$; that is, we demand that $h = 1$ for $n = 0$ thereby obtaining $c = 1$. Note that the above result will imply $\gamma = 9/5 = 1.8$ in the nonlinear regime for any spectra quite *independent* of n! This value of γ – of course - has some observational support.

In principle, the conjecture that $h(n+3)$ is a constant asymptotically can be tested by numerical simulations; but one requires very high dynamic range to obtain a reliable result. A recent analysis[2] of $n = -1, -2$ spectra suggests that $c \approx 1$; that is, $h = 1.5, 3$ asymptoticaly for $n = -1, -2$. This is at best only a consistency check with the limitations of the current simulations. But these simulations *do* suggest that: (i) stable clustering is not a good approximation in an $\Omega = 1$ universe for arbitrary n (ii) h tends to a constant value asymptotically and (iii) the asymptotic limit of h decreases with increasing n.

The above discussion shows that it is indeed possible to make significant progress in understanding the nonlinear gravitational clustering by making a series of physical assumptions. Several possible future directions are suggested by this work. Firstly one would like to tighten the arguments leading to the scaling law in the intermediate regime. Secondly, one would like to understand why h is asymptotically constant (even approximately) for hierarchical models. Finally, one would like to extend these results to understand the behaviour of other physical quantities like velocity dispersion, gravitational potential, higher order correlation functions etc. Such studies - which are in progress - should help one to model quantitatively the flow of power from large to small scales in gravitational clustering at a level comparable to modelling of turbulence in fluid systems.

Acknowledgement

I thank J.S. Bagla, R. Nityananda and J.P. Ostriker for several useful discussions on the ideas described in this work. This work was completed while the author was visiting Department of Astronomy and Astrophysics, PennState University supported by the research grant NAGW-1522.

References

1. See eg., Padmanabhan, T. 1993 "Structure formation in the universe" (Cambridge: Cambridge University Press); chapter 5.
2. Padmanabhan T., Cen R., Ostriker J.P. and Summers, F.J, 1995, Patterns in gravitational clustering: a numerical investigation; astro-ph 9506051; submitted to Ap. J.

3. Peebles, P.J.E. 1980, "The Large-Scale Structure of the Universe" (Princeton: Princeton University Press).
4. Nityananda R., Padmanabhan T., 1994, MNRAS, **271**, 976.
5. Hamilton A.J.S., Kumar P., Lu E., Matthews A., 1991, Ap. J., **374**, L1.
6. Bagla J.S., Padmanabhan T., 1993, IUCAA preprint 22/93; Jour. Astrophys. Astronomy (in press); also see Bagla, J.S. and Padmanabhan, T. 1995, Evolution of gravitational potential in the quasilinear and nonlinear regimes, IUCAA preprint 8/95; astro-ph 9503077.